CREDITS

Publisher: WestNet Learning

Contributing Authors: Tawnya K. Sawyer, David Watts

Managing Editor: Mara G. Gaiser

Book Design: D. Kari Luraas

Book Composition and Index: Amy Casey

Technical Writer and Copy Editor: Larry Beckett

Illustrator: Lynn Siefken

Proofreader: Betty Reed

Cover Design: David Jones

WestNet Learning
Design and Implementation of Voice Networks
Includes illustrations and index

1. Telecommunications Concepts and Components 2. PBX and ACD Systems 3. Point-to-Point Telecommunications Protocols 4. Switched Telecommunications Protocols 5. Traffic Engineering

40-30.0

For instructor-led training, self-paced courses,
turn-key curricula solutions, or more information contact:

WestNet Learning
5420 Ward Road, Suite 150, Arvada, CO 80002 USA
E-mail: Info@westnetlearning.com

To access the WestNet student resource site, go to
http://www.westnetlearning.com/student

Contents

Preface

Although voice networks are evolving from traditional dedicated systems, which operate as entities mostly separated from the data network, to nearly fully integrated components of a converged communications infrastructure, many organizations still maintain and use traditional PBX systems and adjuncts. *Design and Implementation of Voice Networks* describes the rich selection of features available in today's traditional telephone systems, and assists the telecommunications professional in determining which of these features best suit the needs of today's business customers. You will find that many of the features discussed in this course apply to the latest Voice over Internet Protocol (VoIP) systems, as well.

From simple key systems to programmable ACDs, this course describes in detail how telephone systems and adjuncts work to make the business user more productive and profitable. We describe the components and operations typical of customer premises equipment (CPE), and then examine the details of the analog and digital technologies that can link these systems into a wide area voice network. We also introduce IP-enabled PBX systems and present traffic engineering techniques you can use to properly plan a new or upgraded telephone system installation.

Prerequisites

Students will find that basic computer skills, such as using word processing applications, Internet browsers, and e-mail software, are helpful in this course. Students will also find that studies in data networking technologies, telecommunications fundamentals, and Transmission Control Protocol/Internet Protocol (TCP/IP) concepts aid understanding. The following courses covering these topics are available from WestNet Learning:

- Introduction to Networking
- Introduction to Local Area Networks
- Introduction to Wide Area Networks
- Introduction to TCP/IP
- Introduction to Telecommunications
- Internetworking Devices
- Network Design
- Convergence of Technologies

Who Should Study This Course?

This course is intended for telecommunications professionals who want to understand the components and operation of typical customer premises equipment (CPE), as well as the details of the transmission technologies that can link those systems together.

In addition to explaining the basic principles of voice networks, this course is one of the five WestNet Learning courses that will prepare you to study and pass the Avaya Certified Specialist (ACS) certification examinations. By obtaining ACS certification, you validate your skills and knowledge and increase your "marketability" in comparison to your peers.

WestNet Learning's ACS certification series includes the following courses:

- Convergence of Technologies
- Design and Implementation of Voice Networks
- Implementation of Data Networks
- Internetworking Devices
- Network Design

For more information on ACS certification, visit **http://www.avaya-learning.com/** and click on Certification.

Key Topics

- Telecommunications concepts and components as they relate to Local Exchange Carriers, Central Office switching, and trunks, lines, and loops

- Basic telephone services supplied by the Central Office (Centrex service), key systems, and PBXs

- Best practices to use when planning to purchase or upgrade a telephone system, including traffic analysis, trunk requirements, choosing applications, and selecting a vendor

- Fundamental components, features, functions, and configurations of typical business-class PBX and ACD systems

- Detailed descriptions of the key hardware and software components of Avaya's DEFINITY Servers (previously called DEFINITY ECS), Media Servers and Gateways, and the Avaya Call Manager (ACM) call processing application

- In-depth analysis of the capabilities, advantages, and disadvantages of the various analog and digital networking technologies available to tie together geographically dispersed voice systems into a single private or public voice network

- Tools and techniques used to plan and engineer voice network trunk and port requirements and voice mail system capacity, including applying probability formulas to estimate system grade of service, gathering statistical information about existing traffic flows, and planning system size to meet busy hour requirements

Course Objectives

- Describe the process of planning and implementing a telephone network, based on customer requirements and a project plan

- Name the key standards and regulations that affect a telephone implementation project, and explain their relative importance

- Explain some of the practical differences between large voice networks and small phone systems

- Discuss some of the issues to plan for when phones and computers must share the same wiring system

- Explain how the physical design of a voice network can affect its ability to migrate to emerging digital technologies

- Name the project management tools that are required to organize a technology project, and provide examples of how to use each one

- Name the reason for the telecom system's endless "need for speed"

How Do I Use This Course?

This course introduces networking concepts in a series of lessons. Each new concept logically builds on earlier learning. We recommend that you study this course from start to finish, and complete each lesson in order.

A successful network administrator must understand whole systems and fundamental principles. In other words, you must not only remember facts, but also apply them in combinations to solve common network problems. The following sections describe the standard instructional hallmarks found in WestNet study guides.

Real-World Scenarios

In lesson content, review questions, extended activities, and other exercises, this course emphasizes applying networking concepts to real-world scenarios.

Pedagogy

Learning objectives, module summaries, review questions, and extended activities are designed to function as integrated study tools. Learning objectives reflect what you should be able to accomplish after completing each module. Module summaries highlight the key concepts you should master. The review questions help guide critical thinking about those key concepts, and the extended activities provide you with opportunities to practice important techniques.

Key Terms

The information technology (IT) field includes many unique terms that are critical to creating a workable language when it is combined with the world of business. Definitions of key terms are provided in alphabetical order at the beginning of each unit and in a glossary at the end of the book.

Resources Available in Instructor-Led Settings

When this course is used in an academic or instructor-led setting, accompanying instructor resources are often available online. These resources may include some or all of the following: additional exam questions; an instructor's guide with the answers to activities, module quizzes, and the end-of-course exam; and PowerPoint presentations organized by lesson, module, and course.

WestNet Learning's cutting edge administrative tools offer unique, online Windows-based exam software. The online course exam engine includes hundreds of lesson, module, and course questions. These questions can be accessed by individual students and are presented in randomized order, ensuring that students never get the same question in the same order. This feature allows instructors to create printed and online pretests, practice tests, and actual examinations.

Why Choose WestNet?

WestNet Learning offers comprehensive IT educational and certification programs and curricula to secondary schools, colleges, and universities, as well as corporations, resellers, and individual participants around the globe. These programs provide participants with the tools necessary to further their technical knowledge and to obtain hands-on experience. This unique program, which is vendor neutral, helps prepare participants to pursue IT careers, earn secondary and post-secondary educational degrees, and/or obtain industry certification.

WestNet Learning's programs are currently provided to more than 1,000 institutions around the world. Its programs are offered internationally in more than a dozen countries and in five languages.

Introduction

Telecommunications customers have never enjoyed a wider selection of features and technologies than they do today. Small businesses can let a local telephone company handle all of their internal switching, while large corporations can build private voice networks that completely bypass the public telephone system. Customers can choose private branch exchange (PBX) systems that use analog transmission, digital signals, or a combination of the two. In addition, all types of telephone systems offer an impressive level of automation, from small automated attendants that simply answer lines to programmable automatic call distributors (ACDs) that intelligently route calls.

However, this flexibility comes with a price. Telephone technologies are too varied, and too complex, for many customers to understand without help. Customers have never needed the guidance of a telecommunications professional more than they do now.

To make sense of this rich selection of features, and develop telephone systems that meet business needs, telecommunications professionals must think far beyond the walls of a customer's location. The growth of the Internet has required even mid-sized companies to develop global business strategies. When any business telecommunications system can potentially span the planet, a single telephone system can potentially include any communication service offered by any telephone company in the world.

This course provides that broad perspective. We will describe the components and operation of typical customer premises equipment (CPE), then examine details of the digital transmission technologies that can link those systems into wide area voice networks.

Course Overview

Module 1 begins with a review of the structure of the public telephone network, and describes the large networks and transmission facilities that switch telephone calls. It describes the switching systems commonly used by small businesses, and briefly outlines the process of buying a telephone system based on customer requirements.

In Module 2, the focus narrows to PBX switching systems that are essential to most businesses. You will receive a thorough explanation of the components and functions of a typical PBX and ACD, with special emphasis on the architecture of Avaya Communication's DEFINITY Server product line. Additionally, this module discusses the latest Avaya Media Servers and Gateways and Avaya Communications Manager software designed to support enterprise VoIP service.

After reviewing the public telephone network, we introduce the digital transmission services that operate over the network. Module 3 examines point-to-point Physical Layer protocols, such as T1 and Synchronous Optical Network (SONET), while Module 4 focuses on switched Data Link Layer protocols, such as frame relay and Asynchronous Transfer Mode (ATM).

Module 5 concludes by introducing the science of traffic engineering. It describes the three most common methods of estimating the optimum trunk capacity of a telephone system, and offers practical advice for gathering the raw data necessary for traffic engineering calculations.

Module 1
Telecommunications
Concepts and Components

This module reviews the fundamental principles and elements of the public-switched telephone system. We begin by describing the components and functions of a typical local exchange (Lesson 1), and explain the basic operation of the all-important central office (CO) switch (Lesson 2).

We then describe the "products" sold by telecommunications companies: the transmission paths that connect CO switches to each other and to customers. Lesson 3 covers trunks, lines, and loops. We will discuss the difference between analog and digital lines, and see how digital transmission enables a wider range of communications services.

Lessons 4 and 5 focus on two of the most common types of business switching systems, and discuss their relative advantages and disadvantages for different types of organizations. (We will examine the third type of business switch, private branch exchange [PBX], in detail in Module 2.)

Finally, Lesson 6 summarizes common business switching features as it describes a step-by-step process for choosing a telephone system. Although this process focuses on switching systems, its general requirements-based approach is also helpful when shopping for transmission services and other hardware.

Lessons

1. Local Exchange Components
2. Central Office Switching
3. Trunks, Lines, and Loops
4. Central Office Exchange
5. Key Systems
6. Buying a Telephone System

Terms

Advanced Intelligent Network (AIN)—AIN is the PSTN architecture that relies on centralized network servers to make call routing decisions, rather than leaving these decisions up to the individual telephone switches. SS7 control points run protocols that in turn support AIN features, such as caller ID and call blocking.

Automatic Call Distributor (ACD)—An ACD is a programmable system that controls how inbound calls are received, held, delayed, treated, and distributed to call center agents.

Bandwidth—Bandwidth is the total information-carrying capacity of a network or transmission channel. It is the difference between the highest and lowest frequencies that can be transmitted across a transmission line or through a network. Bandwidth is measured in Hz for analog networks and bps for digital networks. See hertz.

Call Detail Report (CDR)—A CDR records on disk or paper the details of incoming and outgoing telephone calls, including source and destination, time of day, and call length.

Caller ID—Caller ID is a service that provides calling party information on a standard telephone line. This information is provided by means of a specified modem protocol between ringing signals.

Central Office (CO)—A CO is the telephone facility where telephone users' lines (local loops) are joined to switching equipment that connects telephone users to each other.

Co-location—A physical and business arrangement to connect the network of a CLEC to that of the ILEC is referred to as co-location. To do this, a CLEC usually installs interconnection equipment at the ILEC's central switching office.

Common Carrier—A common carrier is a company that must offer its services to all customers at the prices and conditions outlined in a public tariff.

Competitive Access Provider (CAP)—A CAP is a company that provides fiber optic links to connect urban business customers to IXCs, bypassing the LEC. As these fiber optic links are placed in major metropolitan areas, CAPs will begin to expand their service offerings.

Competitive Local Exchange Carrier (CLEC)—CLECs are telecommunications resellers, or brokers, who sell data services, Internet access, and local toll calling to businesses and residential customers. Some CLECs route calls over a mix of their own fiber optic, wireless, and copper lines, as well as over facilities they buy/lease at a discount from LECs.

Computer Telephony Integration (CTI)—CTI represents a variety of services made possible by connecting a telephone switch, such as a PBX, to a computer system. For example, a computer can control telephone switching and call routing. Alternately, a telephone switch can pass the identity of incoming callers to the computer system.

Copper Pair—The term copper pair refers to two copper wires that carry voice or data signals to a customer. See local loop.

Direct Inward Dialing (DID)—DID is a process by which a PBX routes calls directly to a particular extension (identified by the last several [usually 4] digits). Incoming trunks must be specifically configured to support DID.

Direct Inward System Access (DISA)—DISA is a PBX feature that allows an outside caller to dial directly into the PBX system, then access the system's features and facilities remotely. DISA is typically used to allow employees to make long-distance calls from home or any remote area, using the company's less expensive long-distance service. To use DISA, an employee calls a special access number (usually toll- free), then enters a short password code.

Divestiture—The breakup of AT&T and the Bell System by the U.S. Justice Department in 1984 is an example of a divestiture. To end an illegal monopoly, AT&T was ordered to separate itself from its 22 local Bell operating companies, which were reorganized into seven RBOCs. AT&T was then restricted to the long distance business, while the RBOCs were limited to local (intraLATA) service. See Regional Bell Operating Company.

E&M—The letters "E&M" are derived from the words ear and mouth; the "mouth" wire of a loop transmits supervisory signals, and the "ear" wire receives them. An E&M trunk uses two pairs of wires (one E pair and one M pair) instead of the single pair used on today's trunks. Some older PBX systems are configured to use two-pair E&M trunks as tie lines (sometimes called tie trunks).

Extension—Voice terminals connected to a PBX/switch by means of telephone lines are referred to as extensions. The term also defines the 3-, 4-, or 5-digit numbers used to identify the voice terminal to the PBX/switch software for call routing purposes.

Fiber Optic—Fiber optic is a thin strand of glass or plastic that transmits a light beam by bending it so that it remains contained within the strand. By using fiber optic transmission, digital signals can travel long distances with a high degree of accuracy.

Foreign Exchange (FX)—FX is a trunk service that lets businesses in one city operate in another city by allowing customers to call a local number. The number is connected, by means of a private line, to a telephone number in a distant city.

Fractional T1 (FT1)—FT1 is a leased-line service that provides data rates from 64 Kbps to 1.544 Mbps, by allowing a user to purchase or lease one or more channels of a T1 link. If a customer needs less bandwidth than 1.544 Mbps, FT1 is a low-cost alternative to choosing a full T1.

Hertz (Hz)—Radio signals are measured in cycles per second, or Hz. One Hz is 1 cycle per second; 1,000 cycles per second is 1 kHz; 1 million cycles per second is 1 MHz.

Incumbent Local Exchange Carrier (ILEC)—An ILEC is the same as a LEC or RBOC.

Integrated Services Digital Network (ISDN)—ISDN is a digital multiplexing technology that can transmit voice, data, and other forms of communication simultaneously over a single local loop. ISDN-BRI provides two "bearer" channels (B channels) of 64 Kbps each, plus one control channel (D channel) of 16 Kbps.

ISDN-PRI is also called T1 service. It offers 23 B channels of 64 Kbps each, plus 1 D channel of 64 Kbps.

Interactive Voice Response (IVR)—IVR is an interface technology that allows outside callers to control a computer application and input information using their telephone keypads. All IVRs can speak back the results of the computer application, and some can also be programmed to fax back the results.

Interexchange Carrier (IXC)—A long distance company (such as AT&T or MCI) that provides telephone and data services between LATAs is referred to as an IXC.

Internet Protocol (IP) Telephony—IP telephony refers to voice telephone service provided over an IP network instead of the public-switched telephone service.

Interoperable—Systems that can work together are considered interoperable. To ensure interoperability, hardware and software manufacturers develop common standards to define the way devices connect and programs exchange information.

Key System—A key system is a simple business telephone system that provides multiple inside extensions access to any of several incoming lines.

Line—A line is a transmission path carried over a physical local loop, but not the same thing as the loop itself. It is a path that connects a telephone switch to an individual user.

Local Access and Transport Area (LATA)—LATAs are geographic calling areas within which an RBOC may provide local and long distance services. LATA boundaries, for the most part, fall within states and do not cross state lines. Each LATA is identified by a unique area code. Calls that begin and end within the same LATA (intraLATA) are generally the sole responsibility of the local telephone company (LEC), while calls that cross to another LATA (interLATA) are passed on to an IXC.

Local Exchange—A geographic region and group of subscribers served by a single CO is referred to as a local exchange.

Local Exchange Carrier (LEC)—A LEC is a company that makes telephone connections to subscribers' homes and businesses, provides telephone services, and collects fees for those services. The terms LEC, ILEC, and RBOC are equivalent.

Local Loop—The pair of copper wires that connects a customer's telephone to the LEC's CO switching system is referred to as the local loop.

Microwave—Microwaves are high-frequency radio waves, commonly used for wireless telephone transmission. While broadcast radio stations usually transmit between 535 and 1,605 kHz, cellular phone systems operate in bands of 824 to 849 MHz and 869 to 894 MHz. See hertz.

Private Branch Exchange (PBX)—A PBX is a sophisticated business telephone system that provides all the switching features of a telephone company's CO switch. Today's PBXs are fully digital, not only offering very sophisticated voice services, such as voice messaging, but also integrating voice and data.

Provisioning—The process of allocating transmission lines, switching capacity, and central programming to provide telecommunications service to a customer is referred to as provisioning.

Regional Bell Operating Company (RBOC)—An RBOC is one of seven companies formed from AT&T's 22 local telephone companies during the breakup of the Bell System. The terms RBOC, LEC, and ILEC are equivalent. The original seven RBOCs were:

- Ameritech
- Bell Atlantic
- Bellsouth
- New York New England Telephone Company (NYNEX)
- Pacific Telesis
- Southwestern Bell Communications
- U S WEST

Request for Proposal (RFP)—An RFP is a formal document that specifies the equipment, software, and services an organization wants to buy, and asks vendors to submit written bids offering their best prices.

Signaling System 7 (SS7)—SS7 (also called SS #7) is an out-of-band system that exchanges control signals and call routing information between CO switches. It is a separate network that connects all COs, regardless of where they are or to whom they belong.

T1, T3—T1 and T3 are two services of a hierarchical system for multiplexing digitized voice signals. The first T-carrier was installed in 1962 by the Bell System. The T-carrier family of systems now includes T1, T1C, T1D, T2, T3, and T4 (and their European counterparts E1, E2, etc.). T1 and its successors were designed to multiplex voice communications. Therefore, T1 was designed such that each channel carries a digitized representation of an analog signal that has a bandwidth of 4,000 Hz. It turns out that 64 Kbps is required to digitize a 4,000-Hz voice signal. Current digitization technology has reduced that requirement to 32 Kbps or less; however, a T-carrier channel is still 64 Kbps. A T1 line offers bandwidth of 1.544 Mbps; a T3 offers 44.736 Mbps.

Tie Line, Tie Trunk—A tie line (tie trunk) is a dedicated circuit that links two points without having to dial a telephone number. Many tie lines provide seamless background connections between business telephone systems.

Transport Carrier—A telecommunications provider or phone company is referred to as a transport carrier.

Trunk—A trunk is a transmission path that connects two telephone switches: two COs, a CO and a business PBX, or two PBXs.

Unbundled Service—Unbundled service refers to a communications channel leased to a CLEC by the ILEC. "Unbundled" means that the ILEC provides only the transmission service, while the CLEC provides management, provisioning, repair, and billing.

Usage Sensitive Pricing (USP)—Usage sensitive pricing is a telecommunications service pricing practice where the carrier bills the customer by line usage rather than by a flat fee. Carriers charge by the call, based on call duration, time of day, and call distance.

Lesson 1—Local Exchange Components

Most people do not care where dial tone comes from, and hardly even notice it, until it is gone. However, it takes a lot of technology and cooperation to provide customers with a dial tone. And, just as the business and legal aspects of the telecommunications industry have gradually evolved over the past 100 years, so has the technology that delivers telecommunications services.

This lesson presents an overview of the structure of the telephone system, and introduces the elements of a typical local telephone exchange. In later lessons, we will learn how these basic components work together to provide service to customers.

Objectives

At the end of this lesson you will be able to:

- Describe the telephone system hierarchy

- Explain what a CO is and what it does

- Describe the purpose of a tandem CO

- Explain the relationship of the customer premises to the public telephone network

 **Key Point**

Many components are necessary to make a telephone call.

The Telephone System Hierarchy

The public-switched telephone network (PSTN) is organized as a multilevel hierarchy. The original telephone system used five numbered levels, as shown on the Original Telecommunications Hierarchy Diagram.

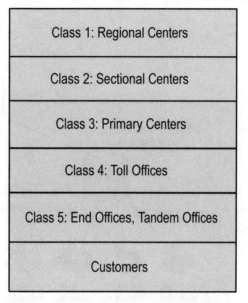

Original Telecommunications Hierarchy

While Class 2 and Class 3 offices are seldom used in today's system, the original numbering system has survived. Therefore, each top-level Class 1 office usually connects to multiple Class 4 offices, skipping the old Classes 2 and 3. Each Class 4 office, in turn, connects to multiple Class 5 offices. The Class 5 offices, or end offices, connect to individual subscribers, as shown on the Current Telecommunications Hierarchy Diagram.

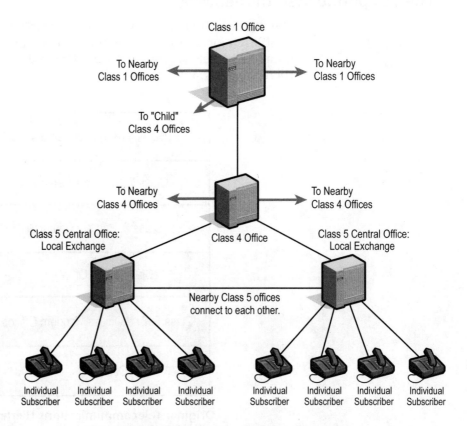

Current Telecommunications Hierarchy

Class 5 Central Office: The Local Exchange

The Class 5 CO is also called the end office or local office. It is the local workhorse for the telephone and data communications traffic in one local exchange. When you pick up your telephone at home, you receive dial tone from a Class 5 CO. There are currently about 1,500 Class 5 local exchanges in the United States.

The Class 5 office is the only office that connects to individual or business subscribers. Offices higher in this hierarchy have only lower level COs as their subscribers. However, each Class 5 CO also connects to other nearby Class 5 offices, as well as its "parent" Class 4 office one level up in the hierarchy.

If a subscriber places a call to another subscriber connected to the same Class 5 office, that office makes the connection directly, as shown on the Call Within the Same Exchange Diagram.

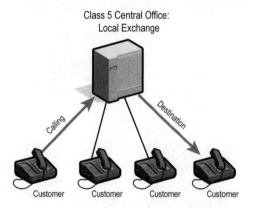

Call Within the Same Exchange

If the caller's Class 5 CO is directly connected to the destination Class 5 CO, the calling CO passes the call directly to the destination CO, which completes the call to the destination subscriber, as shown on the Calling Handoff Diagram.

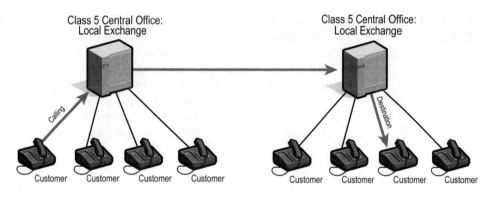

Calling Handoff

However, if the destination CO is not directly connected to the calling CO, or if that connection is too busy, the caller's Class 5 CO passes the call up the hierarchy to its parent Class 4 office, as shown on the Tandem Switching Diagram.

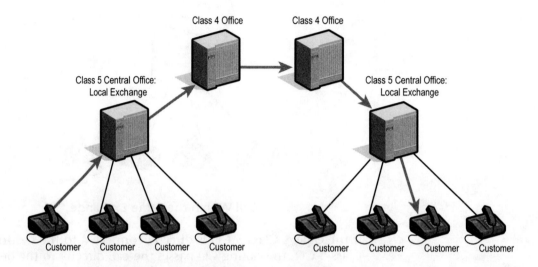

Tandem Switching

Class 4 Central Office

Each Class 4 CO connects to multiple Class 5 offices. Each Class 4 office also connects to nearby Class 4 offices as well as its parent Class 1 office. This interconnection provides alternate paths for calls in the event of a cable outage or, more commonly, congestion on the network. For example, AT&T network engineers can reroute traffic, even down to the telephone number, when an event such as a football game or conference dramatically increases the number of calls to/from a specific location.

AT&T, MCI, Sprint, and other interexchange carriers (IXCs), for the most part, use only Class 4 CO systems. Therefore, we often call these companies Class 4 networks. AT&T's network includes approximately 140 Class 4 CO systems.

Class 4 offices work together in a way similar to Class 5 offices. If two Class 5 COs are both connected to the same Class 4 office, the Class 4 office can switch calls between them. In addition, if the caller's Class 4 CO is directly connected to the destination Class 4 CO, the calling Class 4 CO passes the call directly to the destination Class 4 CO, which passes it to the destination Class 5 CO and the destination subscriber.

However, if the destination Class 4 CO is not directly connected to the calling Class 4 CO, or if that connection is overused, the caller's Class 5 CO passes the call up the hierarchy to its parent Class 1 office, as shown on the Switching Hierarchy Diagram.

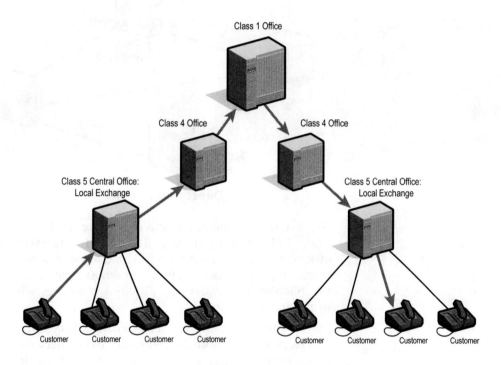

Switching Hierarchy

Tandem Central Offices

As the number of Class 5 COs in a region increases, it becomes impractical to connect every Class 5 CO to every other Class 5 CO. Some COs carry too little traffic between them to justify the cost of a direct connection. Also, the number of connections becomes unmanageable as the number of offices grows, as shown on the Direct Connections Diagram. As you can see from the diagram, six direct connections are needed to link four offices; each office also needs a separate connection to its parent Class 4 office.

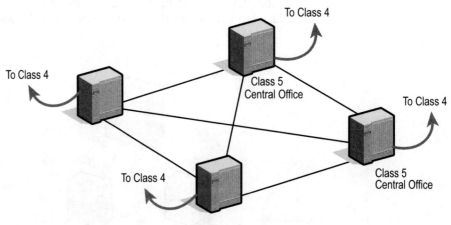

Direct Connections

To solve this problem, tandem offices are used to connect Class 5 COs. Toll tandem offices, also called interexchange tandem offices, connect Class 5 COs to the Class 4 offices of the IXCs. A special type of tandem switch, known as a gateway, interconnects the telephone networks of different countries when their networks are not compatible.

As you can see on the Tandem Central Offices Diagram, tandem offices provide a more efficient way to connect multiple Class 5 COs to one another and IXC switches. Each tandem office connects to other nearby tandem offices and COs, forming a web of interconnections.

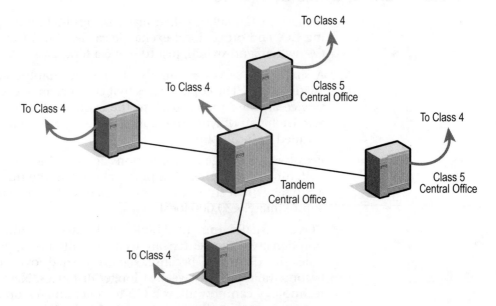

To Class 4

To Class 4

To Class 4

Class 5
Central Office

To Class 4

Class 5
Central Office

Tandem
Central Office

To Class 4

Tandem Central Offices

Tandem COs are considered a type of Class 5 office; however, they do not have direct connections to end users. They serve as intermediaries, switching high volumes of calls between Class 4 and 5 offices.

Class 1 Office

Class 1 offices, or regional toll telephone offices, formed the backbone of the telecommunications system and appear in only a few places in the country. However, purely Class 1 COs are older model switches, and are being replaced by a newer generation of Class 4/Class 5 switches. As these upgrades continue, and Class 1 switches are replaced, the telephone network hierarchy is becoming more "shallow." Rather than reserving the biggest and fastest transmission connections for a few Class 1 offices, telecommunications companies are providing those high-speed connections to their Class 4 COs.

Each of the lower class offices can pass calls up the hierarchy if a connection has failed or is overloaded. However, calls must be handled at the Class 1 level, because there is no higher level to take them. Therefore, a Class 1 office is also known as an "office of last resort."

Class 5 Central Office Components

A Class 5 CO may be called many things including the CO, serving CO, end office, local exchange carrier (LEC) CO, LEC switch, CO switch, and switch, just to name a few.

A single CO serves a roughly circular geographic area, called a "wire center." The copper wires that connect all telephones in this service area are routed back to the centrally located office like a star. Each pair of wires that connects a single telephone to the CO is called a "local loop."

Because of the huge concentrations of wires that converge in local switching offices, there is a practical limit to the number of users a single wire center can serve. Each CO can usually handle approximately 20,000 local loops.

COs are usually unmarked buildings located according to population density. Wire centers are usually limited to approximately 5 miles in diameter, because signals carried over copper wire become unacceptably weak at longer distances. New optical fiber technology can now allow a CO to cover an area up to 150 miles in diameter, including more than 100,000 local loops. However, this more costly technology is usually reserved for rural areas, where subscribers are dispersed over a wider area. In large metropolitan areas, a 5-mile radius remains the norm.

Each CO building contains a Lucent or Northern Telecom (Nortel) CO switching system, backup power systems, connections to other COs, and administrative offices. Multiple computers and associated electronic equipment perform the following key functions of a Class 5 CO:

- Switching and routing calls
- Semipermanent circuit connections
- IXC point of presence (POP)
- Billing

Switching and Routing

The main objective of a CO telecommunications system is to create temporary connections between thousands of pairs of users. Thus, the heart of a Class 5 CO is a piece of complex hardware called a digital CO switch, usually a Lucent 5ESS or Nortel DMS 100. Each of these devices can switch both voice and data connections (keeping the two separate), and provide either Class 5 or tandem switching functions.

Until recently, local switching was a monopoly service offered by the regional Bell operating companies (RBOCs). However, the Telecommunications Act of 1996 opened local switching to competitive local exchange carriers (CLECs). Thus, a CO facility may be owned by an incumbent local exchange carrier (ILEC), or a CLEC.

Semipermanent Circuit Connections

A CO or wire center also establishes dedicated private lines for the exclusive, full-time use of a specific organization. These lines begin and end at fixed points, and are connected within the CO by a device called a Digital Access Cross-Connect System (DACS) or Digital Cross-Connect System (DCS). Each private line can connect two locations of the same organization, a company to a key vendor or customer, or a company directly to another telephone company.

Unlike CO switching systems, which are used to establish multiple temporary connections, a DACS establishes a semipermanent path for voice or data signals. This path is created by a programmed order or administrative action, known as "provisioning." Once created, a path (private line) remains connected until it is disconnected or changed, as shown on the Digital Access Cross-Connect System Diagram.

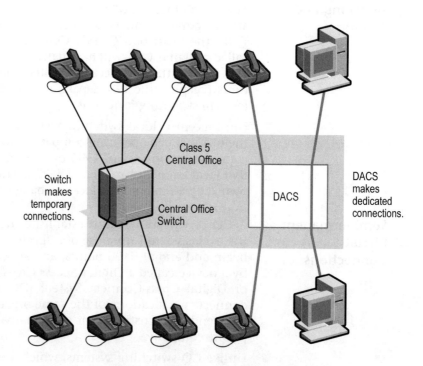

Digital Access Cross-Connect System

The DACS eliminates most of the labor associated with wire rearrangement. All physical wires are attached to the device once, then electronic connections between them are made by instructions to the DACS. This system eliminates extra analog-to-digital conversions, and offers a high degree of flexibility in rerouting circuits. Provisioning changes can also be controlled from a central location over a data link.

IXC POP

If a call needs to be routed outside of a local access and transport area (LATA), it is handed to an IXC for connection to the distant end. The handoff point between the LEC and interexchange carrier point of presence (IXC) is called the carrier's POP. In most cases, a POP is a CO switch, similar to the tandem CO described above. Sometimes an IXC POP is located in the same building as the LEC CO. This arrangement is called co-location, and is quite common.

Billing

Billing is a critical CO function, because all companies that participate in connecting calls expect to be paid for their services. The CO switch creates a Call Detail Report (CDR) that itemizes each call and its duration. This report forms the basis for the final bill the customer receives. Service providers share CDRs to determine the portion of revenues each should receive.

Customer Premises

As we have seen earlier, the telephone hierarchy branches outward like a tree, from the main trunks of the IXCs to the small branches of each local exchange. The "leaves" of each branch are formed by individual telephone customers, located at each "customer premises."

Customer-provided equipment, or customer premises equipment (CPE), originally referred to equipment on a customer premises that was purchased from a vendor other than the local telephone company. Now the term simply refers to telephone equipment (key systems, PBXs, answering machines, etc.) installed on the customer premises.

The point at which the telephone network connects to customer-owned wiring and equipment is called the demarcation point (Dmarc). This is an important definition, because the Dmarc is the point at which the telephone company's responsibility ends, and the customer's begins. Individual customers are responsible for maintaining their own CPE and wiring; the telephone company is only responsible for maintaining the system up to the Dmarc.

Local Exchange Calling Options

The simple telephone hierarchy discussed above is essentially the one the Bell System created a century ago. This is still a good way to visualize the technological system; however, competition has created a more complex set of business arrangements for local calling. Let us explore for a moment some of the ways service can be provided within a local exchange.

LEC Does It All

In the simplest case, an ILEC (probably an RBOC) provides all local service. This was the situation before the introduction of competition.

- The customer is connected to the LEC CO.

- The customer is billed by the LEC.

- The LEC provides all service.

CLEC Serves as Reseller

In this option, a CLEC merely resells a LEC's services. The CLEC bills its customers, or has the LEC print special bills for the CLEC. The CLEC provides customer sales and support.

- The customer's telephone is still connected to LEC CO.

- The LEC bills the CLEC, and the CLEC bills the customer.

- The CLEC provides one-stop shopping: sales, support, and billing.

- The CLEC coordinates services through the LEC.

CLECs Switch and ILECs Wire

In some cases, a CLEC has its own Class 5 CO switch. The CLEC may also have installed some of its own local wiring. However, because the CLEC's wires do not go to all of its customers, the CLEC needs to lease an unbundled loop from the LEC.

The Unbundled Local Loop Diagram shows how this works. The CLEC provides the dial tone to the customer from its own CO switch. The LEC provides a private line from the customer to the CLEC through its CO, using its DACS to cross-connect the line to the CLEC's switch. The CLEC provides switching to the call destination and provides sales/support/billing and other services to the customer.

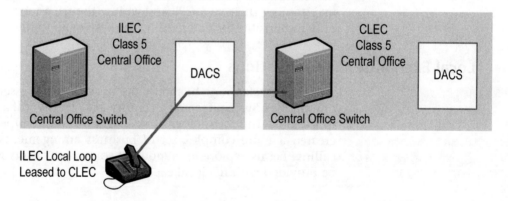

Unbundled Local Loop

CLEC Does It All

A CLEC can also provide end-to-end local service, by providing all of its own switches and loops. In that case, the CLEC and ILEC operate parallel, redundant networks.

However, a CLEC that provides complete local service still must rely on the services of an IXC and distant LEC or CLEC to process many calls. For example, the CLEC may originate the call, which is then connected through an IXC for interLATA calling, and then switched to the destination LEC for completion. In the future, each call may be processed by a number of different carriers, each carrying the data, voice, or video a portion of the way.

Activities

1. Explain how a Class 5 CO routes calls between subscribers when the source and target stations are located on COs not directly connected to each other.

   _____

   _____

   _____

   _____

2. If a Class 4 CO is not directly connected to the target Class 4 CO, how will the originating CO handle the call?

   _____

   _____

   _____

   _____

3. Describe each of the following key Class 5 CO functions:

   a. Switching and routing

   _____

   _____

   b. Semipermanent circuit connections

   _____

   _____

   c. IXC POP

   _____

   _____

   d. Billing

   _____

   _____

4. How do tandem COs differ from standard Class 5 COs?

_____

_____

_____

_____

5. Describe the following ways local exchange services may be provided:

   a. LEC does it all

   _____

   _____

   b. CLEC as reseller

   _____

   _____

   c. CLECs switch, ILECs wire

   _____

   _____

   d. CLEC does it all

   _____

   _____

Extended Activities

1. Contact your LEC or CLEC, and determine what classes of office are local to you. Does your location include Class 1, 4, and 5 offices, or only Class 4 and 5?

2. Does your LEC or CLEC operate tandem offices to interconnect their Class 5 COs? If you have both LECs and CLECs locally, do they share these facilities?

Lesson 2—Central Office Switching

When you pick up a telephone handset in your home or office, a sequence of predefined operations occurs that provides you with a dial tone and the ability to initiate a telephone call. In this lesson, we discuss how a telephone is connected to a communications carrier's CO switch, and how that switch works with other switches to provide a transmission path for voice or data.

It is easier to understand today's telephone technology by showing how it evolved from simple beginnings. Therefore, this lesson begins by explaining some of the earliest telephone switching systems.

Objectives

At the end of this lesson you will be able to:

- Describe the type of switching that has been and is being used

- Explain how a call is made

- Describe the difference between switching and routing

Key Point

The worldwide network of CO switches makes telephone service possible.

The Talking Path

A telephone is a complex electronic device; however, it is based on a simple principle: after an electrical circuit has been established, we can use that flow of electricity to carry a signal. This electrical circuit, between the calling party and called party, is called the "talking path." Just as it was in Alexander Graham Bell's day, a talking path is established at the beginning of a call and maintained for the duration of a call.

Of course, a talking path is not made directly from one telephone to another. Instead, each telephone is connected to the CO switch by a pair of wires. When these wires are connected to each other, a direct-current electrical circuit is completed between the telephone and CO switch. Each telephone contains a switch that completes this connection when the receiver, or handset, is lifted off the switchhook.

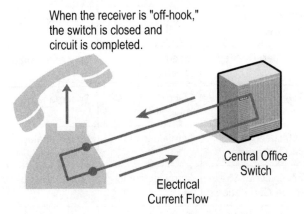

When the receiver is "off-hook," the switch is closed and circuit is completed.

Central Office Switch

Electrical Current Flow

When you complete a call, and place the handset back on its base, the switch breaks the electrical circuit. When electricity cannot flow between the telephone and CO switch, signals cannot travel and the call is ended.

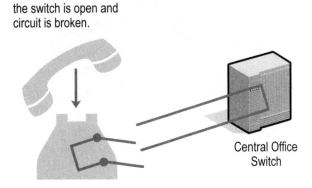

When the receiver is "on-hook," the switch is open and circuit is broken.

Central Office Switch

Evolution of Telephone Switching

COs have gone through a number of fundamental technological changes, as illustrated in the Telephone Switching Evolution Table.

Telephone Switching Evolution

Switching System	Operation	Method of Switching	Type of Control	Type of Network
1878 Manual Operator	Manual	Space/analog	Human	Plug/cord/jack
1892 Step-by-Step	Electro-mechanical	Space/analog	Distributed stage-by-stage	Stepping switch train
1918 Crossbar	Electro-mechanical	Space/analog	Common control	X-bar switching
1960 Lucent ESS – First Generation	Semielectronic	Space/analog	Common control	Reed switch
1972 Lucent ESS – Second Generation	Semielectronic	Space/analog	Stored program control	Reed switch
1976 Lucent ESS – Third Generation	Electronic	Time/digital	Stored program common control	Pulse code modulation

Manual Switching

One hundred years ago, all calls were connected manually. In the late 1800s, each pair of wires from a customer's telephone (each local loop) was terminated in a plug at an operator's console as illustrated on the Tip and Ring Plug Diagram.

Manual Switching System, circa 1880

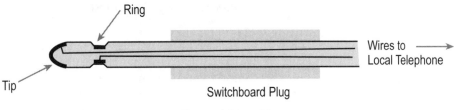

Tip and Ring Plug

Each wire of the local loop pair was connected to a different part of the operator's plug. One wire attached to the tip of the plug, and the other to a ring slightly farther back. Each jack, or socket, of the operator's switchboard contained two metal contacts. When the plug was inserted into the jack, the electrical circuit was completed.

The terms "tip" and "ring" are still used to describe each wire of a local loop, and the analog local loops that provide most "plain old telephone service" (POTS) are commonly called tip and ring circuits.

To place a call, a customer lifted the handset to request service from the CO. When the operator answered, the caller would request the called party by name. The attendant would then plug the caller's wire pair into a horizontal bar line as illustrated on the Operator Plug/Tip and Ring Diagram. He (the first operators were young men) would then yell to the operator who handled the called customer; that operator would connect to the bar and finish setting the call. When the call was completed, another operator would yell to all in the room that the line was clear again.

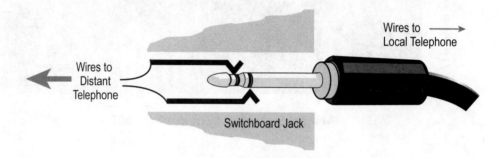

Wires to Local Telephone

Wires to Distant Telephone

Switchboard Jack

Operator Plug/Tip and Ring

Of course, this process was slow, prone to error, and labor intensive. It was only practical when relatively few people had telephones. Some experts estimate that, to manually process today's volume of calls, half the people in the world would have to be telephone operators. Fortunately, someone came up with much better ideas before that was necessary.

Step-by-Step Switching and Rotary-Dial Telephones

The first automated telephone switching was invented in 1891 (or 1893 depending on which reference book you read) by Almon B. Strowger, a funeral director in St. Louis. He was losing business to another funeral company, because the telephone operators were being bribed to manually switch calls to his competitor.

His telephone switch, called a "step-by-step," "stepper," or "Strowger switch," was actually a system that combined a new type of telephone set with an automated switch. To make the two of these work together, each telephone was assigned a unique number.

The new telephone featured a 10-digit rotary dial. To call another party, a caller would first lift the handset to establish the circuit, then dial the digits of the telephone number one at a time. The rotary dial would send a series of pulses, or clicks, down the circuit to the switch: one pulse for the number "1," three pulses for the number "3," and so on.

At the CO, the Strowger switch contained banks of 10-level relays. Each bank of relays represented one of the numbers in a telephone number, as shown on the Step-by-Step Switching Diagram.

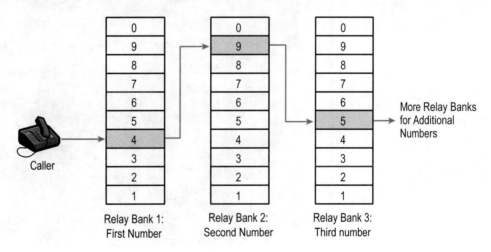

Step-by-Step Switching

Each pulse from the caller's telephone would advance a relay bank by one step. For example, if a customer first dialed a "4," the first relay bank would advance four steps. During the pause between sets of pulses, the next relay bank would be connected, ready for the second digit. When the last digit was dialed and recorded, the combination of the various relay connections formed the talking path to the called party.

Used by permission from the Telecommunications History Group

Step-by-Step Switching System

The step-by-step switch brought greater speed, reliability, and privacy to telephone use. These switches began to be phased out in the 1950s, but a few steppers are still in use today.

Crossbar Switching

The crossbar switch, installed for the first time in 1937 in Brooklyn, New York, used pairs of electromechanical relays (horizontal and vertical) to form a gridwork of potential talking paths. The Crossbar Switching Diagram is a highly simplified description of this complex switch. In the diagram, you can see how contact is made at point A3 by activating the relays for "A" and "3." This common control could be used repeatedly to set up and tear down calls and never sit idle.

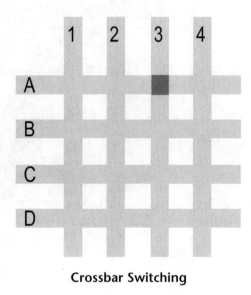

Crossbar Switching

After reaching its peak of installed lines in 1983, the crossbar switch is now largely obsolete because it takes up a lot of space and is not programmable.

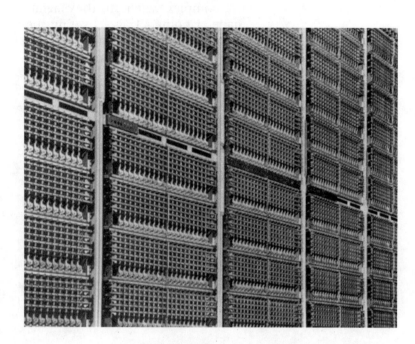

Crossbar Switching System

Later Electro-mechanical Switching

When electronics came along, the electromechanical control of the common control system was replaced with electronics. The network, or switch matrix, was usually replaced with tiny glass-encapsulated reed switches. Only a part of the switch was electronic.

In the next generation, the stored program operation of a digital computer was applied to the switch, although the network remained a complex of reed switches.

Digital Switches

In the next generation of fully digital switches, the talking path through the switch was no longer an electrically continuous circuit. Instead, the sound pattern of speech was digitized into a digital stream of 1s and 0s. However, the local loop between the CO switch and customer remained an electrically continuous analog circuit.

Nortel invented the first all-digital CO switching system in 1979, followed by the Lucent (then AT&T) 4ESS switching system. Today, various types of switches are available for each role in the telephone hierarchy. Most Class 5 switches (central office switches), are one of the Lucent 5ESS family, or the Nortel DMS-100.

Most telephone switches perform the same basic functions, but they do not necessarily work the same way or communicate using the same protocols. However, most switches can emulate the communication protocols used by other switches. For example, most switches can emulate the Lucent 5ESS, which dominates the Class 5 switching market.

Prior to divestiture in 1984, AT&T set telephony standards by means of its research arm, Bell Laboratories. All CO switches, and all lines that carried calls, had to meet prescribed standards. Today's competing hardware manufacturers continue to follow those standards, and cooperate to extend them, so that anyone with a telephone can talk to anyone else. Dialing, ringing, routing, and telephone numbering all conform to uniform standards and plans.

Touch-Tone Signaling

As we saw above, the first step-by-step switches were designed to work with rotary-dialed telephones. Those telephones used "dial pulse" signaling, which produced short, regular interruptions of the direct current flowing between a telephone and switch. The number of interruptions, or pulses, corresponded to the value of the digit. In other words, when you dial the number 5, you hear five clicks.

As CO switching went digital, telephone sets also improved the way they transmitted telephone numbers. The dual tone multifrequency (DTMF) system, commonly called touch tone, uses a pad of 12 buttons. When pressed, each button sends out a combination of two pure tones not found in nature: one high-frequency and one low-frequency. The DTMF Touchpad and Tones Diagram illustrates this concept.

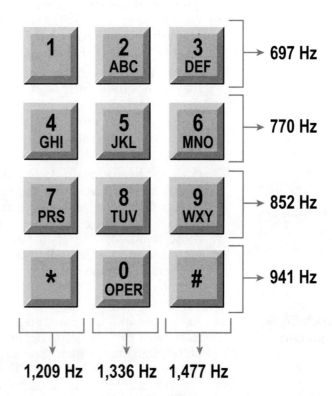

DTMF Touchpad and Tones

By assigning one tone to each row and column, only seven unique tones are needed to identify each of the 12 buttons. These tones can easily be detected by a digital telephone switch or office PBX, a privately owned switching system.

How a Call is Made

When you pick up a telephone handset, a sequence of predefined operations occurs that provides you the ability to use the telephone network. Now that you understand the basic components of the telephone system, let us see how they work together to complete a typical telephone call.

Dial Tone

When you lift the receiver, placing the telephone in the off-hook position, the telephone's internal switch closes the local loop circuit with the CO switch. This allows electrical direct current to flow through the circuit; the presence of this current signals the CO switch that you need a telephone connection. In telephony terms, we say the CO switch has detected the off-hook condition.

The switching module of the CO switch then tests the line and determines its suitability for call processing. If the line tests good, the switch provides dial tone to the caller's telephone.

The off-hook signal also alerts the switch to receive incoming touch tones. If the switch does not receive these tones in a timely manner, it sends a recorded message that reminds the customer that the telephone is off the hook.

Entering a Telephone Number

As soon as the CO switch detects the tones that represent the first digit, it removes dial tone from the line. The switch continues to detect tones and record the corresponding digits, while checking that the number of digits is correct. If the caller enters too few or too many digits, the switch sends the caller error tones or a recorded message.

Call Routing

The switching module of the CO switch then checks with its administrative module to determine the physical transmission path, or routing, the call must take to reach its destination.

If the called party is connected to the same CO switch, the call is connected by that switch. However, the administrative module is advised that those connections are in use, and notes are made for billing purposes.

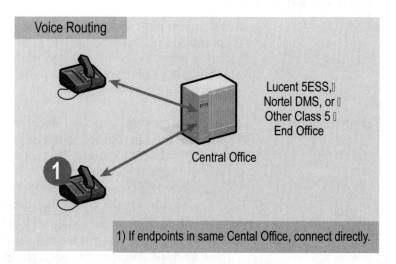

1) If endpoints in same Cental Office, connect directly.

If the called party is connected to a different CO, the call goes through the caller's CO, through a tandem switch, into the called party's CO, then to the called party.

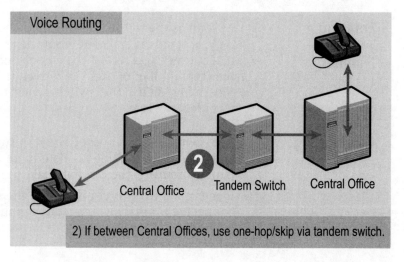

2) If between Central Offices, use one-hop/skip via tandem switch.

What if a particular telephone call is not originated and terminated within the same geographic region? How do we call another city, state, or country? The answer, of course, is to connect the caller's CO to a higher-echelon CO.

Therefore, if a call is not local, it goes through the caller's CO, up to a Class 4 CO (or "toll switch"), into the receiver's local CO, then to the called subscriber.

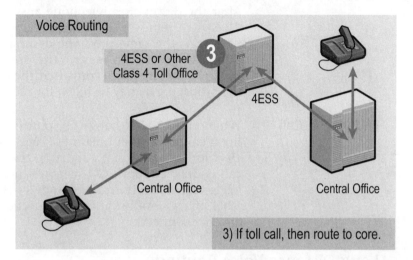

3) If toll call, then route to core.

If the path is blocked at the Class 4 CO, the call is rerouted to another toll switch, into the receiver's local CO, then to the called party (if necessary, the call may be routed up to a Class 1 office).

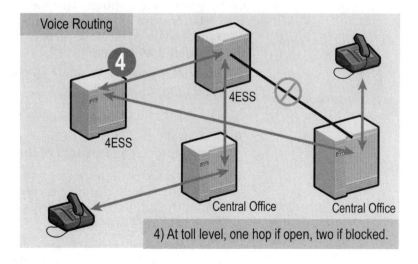

4) At toll level, one hop if open, two if blocked.

Ringing

After a call has been routed to the destination CO switch, the switch tests the line to the called party, to determine whether the line is capable of processing a call. If the line tests good, the switch sends a ringing signal to the destination telephone.

After the destination telephone answers, by going off-hook, the CO switch removes the ringing signal from the destination telephone. At the same time, the destination CO switch signals the calling CO switch to remove the ringing signal the caller hears. Each CO switch also records that the call was completed, so that the proper party may be billed for the call.

Ending the Call

When either caller hangs up, putting one telephone in the on-hook condition that breaks the local loop circuit, the absence of electrical current sends a signal to the nearest CO switch that the call is finished. Both CO switches then perform a series of tests, advise their administrative modules that the call is ended, and label the communications connections as idle and ready for another connection.

Long Distance Voice Routing

This combination of digital switching and touch tone signaling made it simple to introduce direct dialing of long distance and international calls. And, telephone competition is made practical by powerful computers that can track and record the changing relationships between telephone companies and their customers. The typical sequence of steps to switch a long distance call is as follows:

1. The caller lifts the handset and receives dial tone from the local CO switch.

2. The caller enters the called party's telephone number.

3. The local CO switch identifies the call as long distance, based on the number and pattern of the digits.

4. The local CO switch looks up the customer's record to determine which long distance company the customer uses, then routes the call to that company's long distance switch (probably in a Class 4 office).

5. The long distance switch looks up the called party's number to locate the CO switch nearest the called party, and connects to that switch.

6. The destination CO switch tests the line and rings the destination telephone.

7. The called party picks up the handset and begins the conversation.

8. The source and destination CO switches record the completion of the call, and begin to track the duration of the call.

9. Either telephone goes on-hook and the circuit is disconnected.

10. The source and destination CO switches record the final duration of the call for billing purposes, perform testing, and label their connections as idle and ready for another call.

Signaling

As you can see, an important part of the call routing process is the private communications, or signaling, that CO switches use to coordinate the work of setting up and tearing down telephone connections. In general, signaling is the exchange of information between call components required to provide and maintain service.

Signaling means that service-related information is sent between a telephone company and its customers, between components of the same telephone company, and between one telephone company and another. For example, your local CO sends ringing or busy signals to your telephone. When you dial a number, you send an addressing signal to that CO, which then passes the number on to other COs across the country. When you end a call to a distant state, the CO switches that participated in the connection exchange duration information for billing.

In-Band vs. Out-of-Band Signaling

It is important to distinguish between "in-band" and "out-of-band" signaling. In-band signaling shares a single transmission channel with the voice conversation; voice and signaling must take turns using the same transmission path. Analog (POTS) lines use in-band signaling. Therefore, on a POTS line, you can either talk or signal, but you cannot do both simultaneously.

Out-of-band signaling is carried over a separate channel from voice. In other words, it does not take place over the same transmission path as a conversation. For example, Integrated Services Digital Network (ISDN), a popular digital service, uses a separate signaling channel (D-channel) to pass call setup and progression messages. The signaling and call information is carried over the same set of copper wires, but the ISDN protocols build separate frames for each, effectively giving each information type its own share of the overall bandwidth.

SS7

As we have seen, CO switches constantly cooperate to create transmission paths that may span thousands of miles. However, if every CO had to maintain a directory of every telephone number in existence, we might spend hours waiting for a call to be routed. Therefore, telephone COs have their own digital network, entirely separate from the regular telephone network.

This digital network, called Signaling System 7 (SS7), is a very complex system of networks and computers that connects all COs, regardless of where they are or who they belong to. SS7 is a means by which elements of the telephone network, such as CO switches and different telephone companies, exchange signaling and call routing information.

SS7 information is conveyed in the form of out-of-band digital messages that tell each CO switch how and where calls should be sent. SS7 messages also reveal information about each individual call, such as the identity of the calling party. In practice, SS7 creates a dialog among CO switches, as they exchange messages like the following:

- I am forwarding to you a call placed from 212-555-1234 to 718-555-5678. Look for it on trunk 067.

- Someone just dialed 800-555-1212. Where do I route this call?

- The called subscriber for the call on trunk 11 is busy. Release the call and play a busy tone.

- Route 485 is congested. Please do not send any messages to 485 unless they are priority 2 or higher.

- I am taking trunk 143 out of service for maintenance.

Telephone Numbering System

As we have seen thus far, today's heavily used telephone system depends on the ability of callers to place calls without the help of a human operator. That ability, in turn, relies on a system that assigns a unique numbered address to each telephone customer.

However, there is no global standard for telephone numbers. There is a North American standard, shown in the North American Numbering Plan Table, which is used to assign telephone numbers in the United States, Canada, Puerto Rico, and U.S. Virgin Islands, as follows:

North American Numbering Plan

NPA	NXX	XXXX
Number Plan Area (three-digit area code)	Number Network Exchange (three-digit local exchange prefix)	Subscriber Line Identifier (four-digit subscriber identifier)
NXX	NXX	XXXX

In this table, the symbols "N" and "X" indicate the type of number that may appear in a particular position:

- N: any digit from 2 to 9

- X: any digit from 0 to 9

Therefore, as you can see above, an NPA (area code) or NXX (local exchange prefix) may begin with any number from 2 to 9.

However, area codes and local exchange prefixes may not begin with 0 or 1, because these numbers have special meaning to the telephone switch. If the first number the switch receives is a 0, regardless of what number follows, the switch immediately connects the call to the operator. A number 1 in the first position identifies the call as long distance. If a switch receives a leading number 1, it immediately transfers the call to the customer's preferred IXC.

Each Class 5 CO is assigned blocks of NPA-NXXs to distribute to all customers who want telephone service within that wire center. Therefore, all subscribers of the same CO share the same three-digit area code and three-digit local exchange prefix.

The Dwindling Supply of Numbers

Each NPA-NXX block can identify 10,000 unique telephone subscribers, because the four digits of the subscriber line identifier include unique numbers from 0000 to 9999. Following a similar principle, each area code can support approximately 8 million numbers. (Some area codes are set aside for special uses, as we will see below.)

Although this may sound like a lot of numbers, the supply of unique telephone numbers is being exhausted. Urban areas continue to add population, and each individual subscriber now wants multiple lines for fax machines, additional voice services, and Internet access.

Telephone number management within North America is overseen by the North American Numbering Plan Administration (NANPA), which is defined by the FCC. Since 1997, the FCC has selected the company that serves as NANPA through a competitive bidding process. In 1997, NeuStar (then Lockheed Martin IMS) was selected to serve for a five-year term as NANPA. In 2003, NeuStar was again selected to serve an additional five year term that began in July, 2003. Managing all the demands for additional telephone numbers is proving to be a difficult and demanding proposition.

One solution, now being used in large metropolitan areas, is to use multiple area codes, called overlay codes, in the same geographic area. For example, a city's dense downtown district may have a different area code than its outlying suburbs. Or, in heavily populated cities, several area codes may serve the same area. This means residents of those cities must routinely enter 10-digit numbers to make local calls. In addition, the person next door, or even in the apartment below, may have a different area code.

This situation is further complicated by the presence of different telephone companies, such as the new CLECs, who have their own telephone numbers that are different than the LECs. It is even possible that a business subscriber could use one LEC for inbound service, another for outbound service, and still another for Internet access or other services.

LNP

As we saw earlier, the Federal Communications Commission (FCC) has mandated that a long-term solution to local number portability (LNP) be implemented by the telecommunications industry. In other words, the FCC wants customers to be able to switch service from a LEC to a CLEC (or back again) without losing their existing telephone numbers.

This issue is slowly being resolved. However, interim solutions, such as using call forwarding, are in place to allow customers to retain their existing NPA-NXX-XXXXs when changing service providers.

Bell Communications developed Advanced Intelligent Network (AIN) as a model technology designed to support advanced call routing services over the SS7 network. Vendors who offer AIN services advertise that it will allow users to change local telephone companies without having to change their telephone numbers, thus enabling LNP. The creation and implementation of a standardized number portability plan will be a tremendous relief for both service providers and consumers. However, this begs the question, "Who owns a telephone number?" Someday will telephone numbers, like Social Security numbers, be assigned at birth?

As of this writing, LNP is being applied to wireless phone numbers when moving from one wireless carrier to another. It has been implemented in some areas and is slated to be universally available in the United States by May 2004. The status of wireline-to-wireline number portability is currently unclear.

Assigning Telephone Numbers

Currently, most telephone companies assign telephone numbers in three ways:

- Random assignment, or "what you get is what you get," is the most common method of assigning numbers. The telephone company simply assigns the next number from a pool of available numbers in the customer's local exchange.

- Special-request numbers are also available for an extra fee, providing easy-to-remember business numbers such as 444-9000 or 444-TAXI.

- Numbers may also be reserved for future use. However, no LEC guarantees a number assignment until it is actually installed.

Special Area Codes

Digital switching made it possible to use the telephone numbering system to access special services instead of geographical areas. A brief overview of some of the most popular services available through special-purpose NPA (area) codes is presented below.

Special NXX Codes

NXX codes usually identify CO; however, some NXXs, with or without a special NPA, are set side to access special services as described below.

Special Information: 555

555 numbers access special information services, such as long distance directory assistance. The line number (XXXX) identifies each individual service.

**Hearing-
Impaired
Services: 800-855**

800-855 numbers, in the format 800-855-XXXX, provide free access to statewide relay services, such as Telecommunications Relay Service (TRS) and Message Relay Service, which provide trained assistants that translate calls between the voice telephones of hearing customers and the teletypewriters used by hearing-impaired customers.

Service Codes, or N11 Numbers

Like speed-dial numbers, service codes are three-digit numbers that directly connect customers to local exchange special services. They are commonly called "N11" numbers because of their numbering format, which follows the same rules as NPAs and NXXs. In the United States, the FCC administers N11 numbers, which include:

*   **211**—Community Information and Referral Services (United States)

*   **311**—Nonemergency Police and Other Governmental Services (United States)

*   **411**—Local Directory Assistance (sometimes 1411)

*   **511**—Traffic and Transportation Information (United States), Reserved (Canada)

*   **611**—Repair Service

*   **711**—TRS

*   **811**—Business Office

*   **911**—Emergency

**Toll-Free: 800,
888, 877**

Toll-free calls (the called company, not the caller, pays for the call) have been available for more than 20 years, and consumers have come to expect companies to provide them as a customer service feature. With such a great demand for these calls, the 800 NPA code has run out of available numbers. Therefore, additional NPA codes, such as 888, 877, and 866, are now used for toll-free calls.

Some companies use toll-free numbers to dial into their PBX systems, which then gives them access to special outgoing long distance lines. This PBX system feature is called Direct Inward System Access (DISA). Although DISA can often be a cost-effective system for providing long distance service for traveling employees, hackers have attacked some companies and used these lines to steal long distance service.

Premium: 900

Unlike 800 numbers, which companies provide free as marketing and customer service tools, a 900 number is often used as a revenue-generating product. When calling a 900 number, the customer not only pays for the long distance charge, but is also charged a premium by the called company. The extra fee can be as high as $50 per minute.

Companies are using 900 numbers for customer support, fund raising, and pay-as-you-go services. By using a 900 number, a company can simplify its billing procedure. Because the customer's telephone number is billed, the company does not need to create credit card transactions or issue a bill.

International: 700

700 numbers are used for international data calls at 56 kilobits per second (Kbps). The use of 700 numbers is diminishing as ISDN, also an international standard, becomes more widely used for data transmission.

Follow-You: 500

One of the most exciting emerging types of calling services is "follow-you" calling. By dialing one 500 number, callers can arrange to forward calls to their current location from home, car, or pager.

Activities

1. Describe DTMF.

   _____

   _____

2. Describe the sequence of operations that occur when a call is placed.

   _____

   _____

   _____

3. The type of telephone switch that uses pairs of electrome-chanical relays to form potential talking paths is a _____.

   a. Step-by-step switch

   b. Crossbar switch

   c. Digital switch

   d. Manual switch

4. A _____ uses pulses from the subscriber's telephone set to advance relay banks, forming a talking path to the called party.

   a. Step-by-step switch

   b. Crossbar switch

   c. Digital switch

   d. Manual switch

5. Nortel invented the first of these switches in 1979:

   a. Step-by-step switch

   b. Crossbar switch

   c. Digital switch

   d. Manual switch

6. Describe the difference between switching and routing in telecommunications.

   _____

   _____

Extended Activities

1. Contact your LEC or CLEC and inquire which of the discussed switch types they use to complete calls. Do you use tone or pulse dialing in your area?

2. Visit **http://www.avaya.com** to review their CO products.

Lesson 3—Trunks, Lines, and Loops

As we saw in Lesson 2, today's telephone system is made possible by a network of sophisticated digital switches and computers. However, those devices need some way to communicate with each other, and with the customers they ultimately serve. In other words, they must be connected in some way.

The telephone network is physically connected with a combination of copper wire, optical fiber, television coaxial cable, and wireless radio systems. Various types of physical connections can support different styles of signal transmission. In other words, you can use the same wire, optical fiber, or cable to move information in several different ways. The transmission methods, not the wires or optical fiber that carry them, are what telecommunications businesses sell. In this lesson, we will review the "products" of the telecommunications business: the main methods of transmitting information across a connection.

Objectives

At the end of this lesson you will be able to:

- Explain the difference between a trunk and a line

- Describe the most common types of local loops

- Explain the difference between analog and digital signals

 **Key Point**

A loop is like a road that can carry different types of traffic. The type, or size, of a road can influence the type or volume of traffic the road can carry.

Telecommunications Terms

The telecommunications industry uses several special terms to define and classify its transmission services. Some of these terms are used loosely; thus, we will first spend some time clarifying their meaning. These terms describe the various means by which telecommunications circuits are connected. The Telecommunications Network Diagram illustrates these.

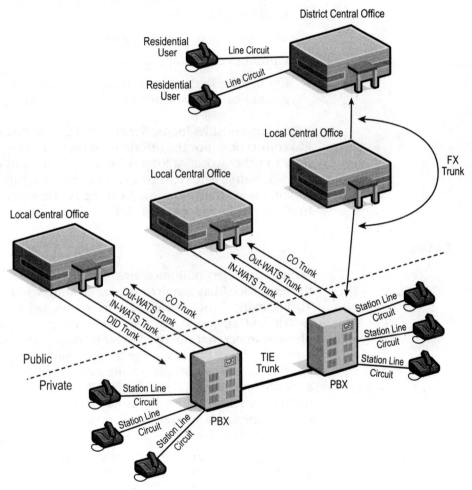

Telecommunications Network

Loops

A loop, or local loop, is the physical connection that links the Dmarc at the customer premises to the telephone company CO switch. Traditionally, and most commonly, the local loop is a twisted pair of copper wires between an individual telephone set and the CO switch, or multiple twisted pairs between a business switching system and the CO switch. However, local loops can take a wide variety of forms:

- Copper twisted pair (up to 45 megabits [Mb]) residential/business

- Television cable (coaxial cable)

- Optical fiber

- Cellular/Personal Communications Service (PCS), using ground-based or satellite radio transmission

When you consider loops, remember the term describes the physical connection, not the information that travels over the connection. In other words, a loop is like a road that can carry different types of traffic. The type, or size, of a road can influence the type or volume of traffic the road can carry. However, the road and traffic are still two separate things.

Lines

During the telecommunications system's 100-year development, the term "line" has acquired several meanings and interpretations. Traditionally, "line" means a connection between a telephone switch and an individual telephone, usually an individual telephone number that can be used for incoming and outgoing calls. A line, as in this definition, is the most common type of loop. Generally speaking, when people use the word "line" by itself, they mean a POTS local loop. For example, if your neighbor says, "We installed a second line," he probably means the telephone company connected a second pair of wires to his home. Thus, in this case, the terms "line" and "local loop" are interchangeable.

However, a line can be more generally understood as a transmission path that connects a CO to an individual customer, or a company's PBX to a single employee's telephone extension. In this sense, a line describes the traffic that travels over a loop, as well as the volume and speed of that traffic. For example, we can refer to slow-speed lines such as a fire alarm circuit, or high-speed lines such as a fiber optic data transmission link.

If you think of a local loop as a roadway, as we did above, you can think of a line as one traffic lane on the roadway. The larger the capacity of the loop (road), the more lines (traffic lanes) it can carry. Thus, you can imagine the copper local loop to your home telephone as a narrow, one-lane gravel road. Under normal conditions, the loop can only accommodate one transmission (voice or data) at a time, and at fairly slow speeds. In contrast, a fiber optic cable is a high-speed, multilane highway. A single optical loop can support multiple simultaneous, high-speed transmission paths. Most important, each path can be considered, and sold to customers, as a separate line.

In this course, we use the term "line" in this second, more flexible, sense. We refer to a line as a transmission path carried over a physical loop, but not the loop itself.

Trunks

Like the term "line," "trunk" is often used interchangeably with "loop." However, like "line," this course uses "trunk" to describe a transmission path, not the physical medium that carries the transmission.

A trunk specifically describes the connection between switching systems. For example, a trunk connects two COs, a CO and business PBX, or two PBXs.

Trunks can be classified according to three major characteristics:

- **Direction**—Incoming, outgoing, or two-way traffic
- **Capacity**—Quantity of information the trunk can carry
- **Transmission type**—Analog or digital

Trunk Direction

Trunks are offered to business customers in three configurations. Incoming trunks carry calls from a CO switch to a PBX. Outgoing trunks carry calls from a PBX to a CO. Two-way trunks can be used for both making and receiving calls.

Trunk Capacity

Capacity means the information-carrying ability of a telecommunications facility: the loops, switches, and other hardware that form a transmission path. Trunks usually have higher capacities than single lines, because of the larger volumes of traffic they usually carry. A trunk may be provisioned over a bundle of twisted pair wires, or a single optical fiber, but it often provides multiple-line service.

The capacity of a trunk can be measured by the number of individual lines it provides, or, in the case of a digital data trunk, its transmission speed in bits per second (bps). Trunk capacity is also expressed as the number of channels the trunk can carry. However, before we can discuss the concept of channels, we must first explain the difference between analog and digital transmission.

Analog Transmission

"Analog" means a signal is carried as a pattern of continually changing waves. The first telephone system was analog, and analog transmission is still used on most POTS voice circuits to and from homes.

The sounds you hear are caused by waves carried through air, as our old friend Alexander Graham Bell discovered. The microphone on a telephone can convert these sound waves into a similar pattern of electromagnetic waves. The electromagnetic waves are carried across the telephone network to the speaker on a distant telephone. The speaker converts the electromagnetic analog waves back into sound waves, making the signal audible to the human ear.

There are only two basic qualities of an analog signal:

- Frequency refers to the number of times per second a wave swings back and forth in a cycle from its beginning point to its ending point. Think of frequency as the number of wave crests, or cycles, that passes a fixed point during a particular period of time. Therefore, frequency is measured in cycles per second, or hertz (Hz). The frequency of a sound (air wave) signal determines its pitch. On the Frequency Diagram, the high-frequency (closely spaced) waves would create a high-pitched sound, and the low-frequency (loosely spaced) waves would create a low-pitched sound.

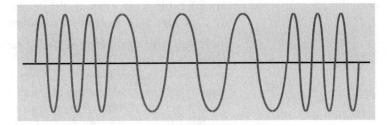

Frequency

- Amplitude refers to the height of a wave, or how far from the center the wave swings. Generally speaking, amplitude describes the power, or loudness, of a signal. On the Amplitude Diagram, the frequency does not change, but the amplitude does. Therefore, the pitch of the sound does not change, but the loudness changes from soft to loud, then back to soft again.

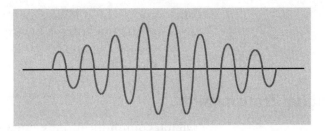

Amplitude

Thus, frequency determines the pitch of a sound, and amplitude determines the loudness. By varying these characteristics, engineers can record or play back almost any sound.

Channels and Bandwidth

To get back to the idea of channels and trunks, think of a radio. Each radio station transmits its signal using a narrow slice of the overall broadcast radio spectrum. Radio stations can do this because a small slice of frequencies (4,000 Hz difference between the highest and lowest) is all that is necessary to carry the wave patterns that recreate speech.

Now, as we explained earlier, some types of physical transmission media can carry more information than others. Some, like a POTS local loop, are normally limited to only one channel. Therefore, you can use your home telephone line to talk, or surf the

Internet, but not simultaneously. Others, like a coaxial (cable television) cable, can carry multiple simultaneous signals by assigning each signal to a range of frequencies, just as a radio does.

Therefore, the more frequencies a medium can handle, the more channels of information it can carry. Each assigned slice of frequencies is called a band. Therefore, the information-carrying capacity of an analog transmission path is called its bandwidth. The copper local loop to your home telephone has a very narrow bandwidth, only enough for one voice channel. In contrast, businesses require trunks with much wider bandwidth, often 24 channels or more.

A channel is a basic building block of business telephone service. While telephone companies rent entire multichannel trunks to businesses, they also provide individual channels from those trunks. One business might lease a high-capacity trunk, then use special equipment to split its bandwidth into many smaller channels. Another business might lease several individual channels, then use equipment to combine them into one mid-sized transmission path.

This process of dividing or combining individual channels is accomplished by using a device called a multiplexer (MUX). Multiplexing is a digital process, so let us see how digital signals are different from analog.

Digital Transmission

Digital communication is a newer technology that forms the basis for today's telecommunications networks, services, and systems. Digital technology is now being used to carry many types of signals (voice, data, video, and more) on the same physical media, and often at the same time.

Instead of a pattern of continually changing waves, digital signals are transmitted in form of binary bits: information represented as a series of 1s and 0s. The term "binary" refers to the fact that there are only two values for a bit: on or off. "On" bits are depicted as 1 and "off" bits are depicted as 0. When bits are transmitted over wires, a 0 is represented by the absence of electricity, and a 1 is represented by the presence of electricity.

There are two reasons why digital technologies create more accurate and clearer voice communication. First, digital bits can only be on or off. In other words, it is more complex to recreate an analog wave that can have multiple complex forms, than a bit that can only take two forms.

Second, digital transmission makes it easier to distinguish between signal and distortion, or "noise." In analog transmission, both the signal and any accumulated noise are transmitted together, and amplified along the way. Because both speech and line noise are represented as complex wave forms, it is very difficult for the transmission system to tell the difference between the real signal and any noise that has distorted the signal.

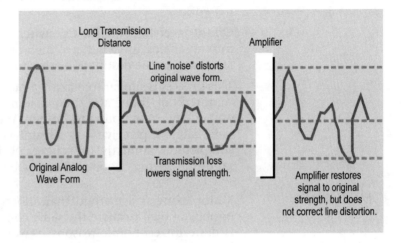

Digital signals are also susceptible to distortion; however, they are more reliable because electronic equipment can more easily recognize the original pattern of high and low electrical current.

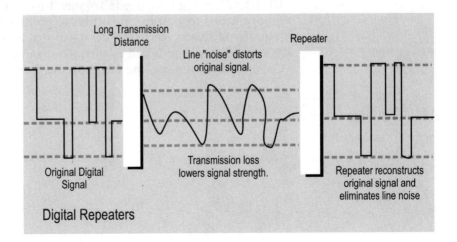

63

PCM

The analog vibrations that represent the sound of your voice are converted into digital electrical impulses through a technique called pulse code modulation (PCM).

PCM is a sampling technique. The analog signal that represents a voice is measured, or "sampled," 8,000 times per second. Each of the samples is converted into an 8-bit code that represents the frequency and amplitude of that sample of sound. The resulting stream of binary bits is then transmitted across the telephone network.

On the receiving end, the CO switch converts the binary bits back into an analog wave pattern, then transmits the pattern over the local loop to the destination telephone.

This process is similar to making a movie. The action of a movie is captured one frame at a time at some number of frames per second, a kind of sampling technique. When it is played back at the same rate, we perceive it as a seamless action, and not a series of still pictures. Similarly, users never notice that their voice conversations are being chopped into tiny bits of data.

Multiplexing

Multiplexing is a method that allows multiple digital messages (groups of bits) to share the same communications channel. This is done for economic reasons. Fewer channels mean less wires, less equipment for connection of the wires, and less overhead, maintenance, and so on.

To understand multiplexing, imagine three groups of people lined up to board the same escalator, as illustrated on the Multiplexing Diagram. Each group needs to stay together, but none of the groups wants to wait for the others to go first. To solve this problem, the MUX (an electronic device in real life) allows one person from each group to board in turn. For example, one person from Group A will step on, then one person from Group B, then one from Group C. Then the pattern will repeat with another person from each group, until all have boarded.

Multiplexing

At the top of the escalator, a demultiplexer sorts the people back into their original groups. If the escalator is fast enough, all three groups will make smooth progress, and none will notice any delay during the multiplexed part of their trip.

There are two dominant types of multiplexing:

- **Time-division multiplexing (TDM)**—The MUX assigns each signal a slice of time to use the transmission path.

- **Frequency-division multiplexing (FDM)**—The MUX assigns different signals to different frequencies, just as a radio does.

Each time slot or frequency range can form a separate channel. The problem with both TDM and FDM is that a channel is provided to each caller or user whether they use it or do not say anything at all. For example, if a signal has no bits to send when its assigned time comes around, then the time slot remains empty and is wasted, just as a reserved seat on a train or plane is wasted if a traveler does not arrive on time. There are more complex multiplexing technologies that solve this problem, but they are beyond the scope of this book.

**Digital
Bandwidth**

The information-carrying capacity of a digital line or trunk is expressed as the number of bits it can transmit per second. Digital bandwidth is commonly measured in Kbps and megabits per second (Mbps). One kilobit (Kb) is approximately 1,000 bits; and 1 Mb is approximately 1 million bits.

Special Access Trunks

Thus far, we have described trunks that provide multiple voice channels to businesses. However, trunks can come in a variety of capacities, and provide various types of service. Some of the more common special uses for trunks are described below.

Tie Lines

A tie line is a dedicated circuit, connected through a CO DACS that links two points without having to dial a telephone number. A tie line may be accessed by lifting a telephone handset, or pushing a few buttons. Many tie lines provide seamless background connections between different business systems. For example, companies that have multiple PBX systems at two or more locations often use tie lines to connect the PBXs. Because the connections are always "on," the different PBX systems can be configured to behave as one large system. The Tie Trunk Connections Diagram illustrates PBXs located in different area codes, connected via Tie Trunks.

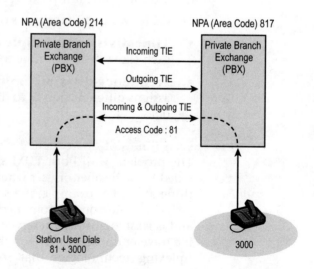

Tie Trunk Connections

Incoming, outgoing, and bi-directional tie trunks connect the PBXs. The carrier's network provides one-way trunks to handle inbound and outbound PSTN calls, and two-way trunks connect PBX users in the remote sites. PBX users dial an access code to access the two-way trunks, and the CO switch connects the two over the dedicated tie trunks.

Private Lines

A private line creates an end-to-end, on-all-the-time connection between two locations. No LEC or IXC switch routes or bills per call; each private line is connected through a CO DACS. A customer has unlimited use of the private line for a fixed monthly rate. Private lines may be provisioned within a LATA, between LATAs (interLATA), or even internationally.

Private lines are more popular for data applications than voice, but may be used for either. Depending on their needs and budgets, customers may lease a single 64-Kbps channel, full T-1 service (with 24 channels of 64 Kbps each), or even higher bandwidth services. Private lines are typically billed a flat rate per month based on the amount of bandwidth leased and number of airline miles between the two connection points.

FX

Foreign exchange (FX), not to be confused with FAX, refers to a foreign exchange trunk. In this case, "foreign" means a nonlocal CO, not a location outside the country.

Companies use FX trunks to provide local numbers in cities where the companies do not have offices. Customers can call the company toll free, and may not even be aware the company is located in a different city. FX trunks are also used to transfer data between two distant locations, because they provide high bandwidth and security.

As shown in the FX Diagram, an FX trunk starts at the customer's location, connects to the local CO DACS, and extends from there to another foreign CO anywhere in the country. There is a fixed monthly charge for all mileage; however, there are no usage-sensitive charges for the miles. The remote caller dials an access code that signals the local PBX to seize the FX trunk to the CO. The CO switch recognizes that the call is an inbound FX call, and in turn seizes the trunk to the remote PBX. The CO switch passes the call to the foreign exchange PBX, which then rings the destination telephone set.

Though the call is placed to a remote user in another area, nonetheless the local caller only needs dial the access code and a seven digit number, which appears as a local number on the remote PBX.

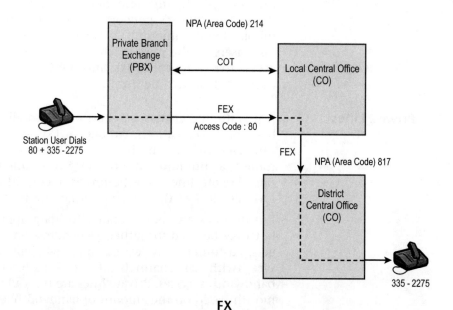

FX

Ring Down

A ring down trunk is a special telephone line that rings a particular destination telephone as soon as the caller picks up the handset. It is a type of dedicated line that permanently connects two telephones; a ring down trunk cannot be used to call other numbers.

Ring down trunks are used to connect one station directly to another. For example, airports commonly feature courtesy telephones from car rental agencies or hotels. When a customer picks up the handset, one of these telephones instantly rings its central reservations desk.

Trunk Pricing

Depending on a company's need for speed or bandwidth, trunks are acquired based on how much they are used. For the most part, companies get flat-rate trunks; the company pays a fixed monthly fee and can use the trunk as much as necessary. However, some companies only need to use a trunk part-time, thus trunks are also billed as a measured service, also known as usage sensitive pricing (USP). Here, customers only pay for the service they use; carriers say that this saves low volume subscribers money. And, of course, for high-volume or large-usage customers, a variety of discounts are always available.

Activities

1. Describe the difference between a line, loop, and trunk.

   _____

   _____

   _____

2. List some forms local loops can take.

   _____

   _____

3. How does analog transmission differ from digital transmission?

   _____

   _____

   _____

   _____

4. A sampling technique used to represent an analog wave form as a digital code is:

   a. PCM

   b. Multiplexing

   c. TDM

   d. FDM

5. A trunk type that creates an end-to-end, always-on connection between locations most often used for data transmission is a(n) _____.

   a. Tie line

   b. Ring down trunk

   c. FX trunk

   d. Private line

6. What is a type of multiplexing that assigns digital signals a slice of time?

    a. PCM

    b. FDM

    c. TDM

    d. POTS

Extended Activities

1. Explore the various trunk types available from your LEC and/ or CLEC. How do their rates differ by service, bandwidth, and distance?

2. In your area, what options are available for the local loop? Is fiber access to the CO available? Do any local service providers make wireless data services available? If so, on what type of wireless technology are these based?

Lesson 4—Central Office Exchange

When a business reaches 18 to 20 lines on its key system, the telephone sets become so large and complicated that other solutions are usually needed. Many businesses with key systems need the features only available on PBX systems. However, many small and mid-sized businesses do not want to commit the money, personnel, or space to their own PBX systems. They want nothing to do with owning or managing a telephone system.

LECs offer an economical solution: system features normally associated with PBX systems can be leased through a LEC and provided through a CO switch. Changes and upgrades are easily implemented through the LEC. This solution, called Centrex (for Central Office Exchange), can be used in either analog or digital lines/trunks.

Objectives

At the end of this lesson you will be able to:

*   Describe how the Centrex service works

*   Explain the types of businesses that would benefit most from Centrex

*   Name the main reasons for choosing Centrex over a PBX

Key Point

Centrex is good for businesses that have multiple locations.

Centrex

Centrex is a telephone company network service that provides sophisticated office telephone switching features through a CO switch. Some LECs use other names for the same service, including Centranet, Essex, and Plexar.

In a Centrex system, a business customer purchases a block of telephone numbers. The LEC assigns each of these numbers to one of the customer's telephone lines and telephone sets. In effect, this directly connects each customer telephone to the CO switch, which provides office switching features and intercom services among the lines.

The most likely Centrex customers are businesses and organizations that have several locations across one area of a city. For example, a city government, bank with branch offices, or one-business campus are all likely Centrex customers. These organizations find it convenient to tie their dispersed locations together with centralized switching services provided at a CO, as illustrated on the Multiple Locations Linked by Centrex Diagram.

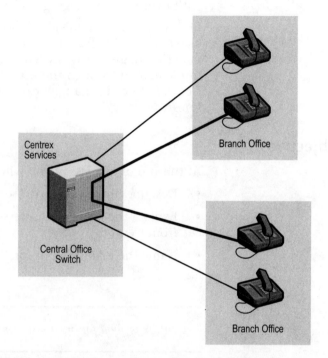

Multiple Locations Linked by Centrex

With Centrex, these dispersed locations can perform as a single telephone system. Employees at all locations can use three-, four-, or five-digit extension numbers to talk with each other, without having to dial 9 to get an outside trunk line.

Centrex Benefits

Following divestiture in 1984, the LECs began aggressively marketing Centrex to the less than 100-line market. This business market is the fastest growing, with the largest number of sites; thus, it constitutes the largest portion of the on-site telephone equipment market. Centrex offers real benefits to these businesses, if they need advanced switching features but do not want to manage their own telephone system. LEC marketing campaigns stressed the following advantages:

- **Reliability**—Redundant features are built into CO switches, making it less likely they will fail.

- **Space savings**—Centrex equipment is located in the CO, not the business.

- **Ease of administration**—The telephone company takes care of maintenance and programming.

- **Ease of upgrades**—The CO can easily keep up with business growth.

- **Compatibility**—Centrex is compatible with new telephone network technologies.

- **Less hardware expense**—No on-site switching means lower power and cabling expenses.

- **Vendor support**—The telephone company will never go out of business.

Technical Requirements for Centrex

There are several technical requirements for implementing a Centrex system, as discussed in this section.

Distance From CO Switch

Because Centrex switching is provided by a CO switch, all of a company's locations or offices must be located within the local exchange(s) controlled by a single LEC. The exchanges may be scattered across a wide area, as long as all are controlled by the same LEC. The telephone company connects the various lines through the DACS in each CO, leading them back to one central CO that provides Centrex switching. The carrier can also program the various local CO switches to work together, providing seamless service across a city or region.

As customers grow, Centrex is able to accommodate large city-wide situations, as well as having a mix of long distance or interexchange services. One of the challenges faced in managing Centrex is the mileage charges associated with hundreds or thousands of lines between locations. In some cases, companies with a vast number of locations over a wide area require more than one CO.

| **Allocation of Call Paths** | Centrex customers must determine the number of incoming and outgoing lines or call paths, just as they would to implement a key system. In a Centrex system, individual lines are also known as Network Access Registers (NARs). We need a NAR for every simultaneous incoming or outgoing call path we anticipate we will need. |

| **CPE** | A business customer must provide its own telephone sets and attendant consoles. This customer premises equipment (CPE) is connected, usually by means of twisted pair loops, to a part of the LEC CO called a common block, distribution frame, or main distribution frame. All three of these terms refer to the part of the CO where the end of each local loop is connected to a "common block" of connectors that make it simpler to attach the loops to the CO switch. |

The CO common block determines the features, restrictions, and access that each Centrex customer may receive from that CO. Depending on the capability and sophistication of a particular CO, Centrex service may be provided by its own separate CO switch. In other COs, Centrex may be provided by software functions within the main CO switch, which includes slots for programmable electronic circuit boards.

Centrex Features

Centrex makes available all features and functions of a CO switch. A few of the more popular features are:

- Voice mail
- Conferencing
- Call forwarding and special routing
- Call waiting
- Speed dialing
- Call hold/consultation hold
- Call pickup
- Call transfer
- Distinctive ring

The telephone companies have, over the years, continued to add features to Centrex to make it competitive with PBX systems, including tie lines to other Centrex or PBX systems.

Although Centrex provides a wide array of features, many customers use PBX systems, key systems, and Centrex together. This is because customers who have many locations may also have a large number of telephones at these locations. For example, each of several large offices might use a PBX for intraoffice switching, and Centrex for interoffice traffic.

Centrex vs. PBX

Despite the many advantages Centrex has over PBX systems, customers have tended to prefer PBX systems for the last 20 years. This trend may change as CLECs and LECs become more competitive and Centrex prices fall.

The Centrex and PBX Comparison Table summarizes the main reasons a customer would choose either Centrex or a PBX system.

Centrex and PBX Comparison

Centrex	PBX
Off-site telephone company switching	On-site switching
Service to Dmarc	Service to telephone
Telephone company staff	Company staff
Telephone company space, power, HVAC	Company space, power, HVAC
Rent/no property tax	Lease/buy
Linear growth (gradual upgrades)	Step growth (larger upgrades)
Multilocation	Single location
Complicated billing	"Locked-in" pricing

The point of the table is that one solution is not right for all customers. For many years, customers preferred PBX systems. Today, with more alternatives to local service available, customers are giving Centrex another look.

Activities

1. LEC marketing stresses which of the following Centrex advantages?

    a. Centrex equipment is located on the customer premises.

    b. Centrex equipment and customer site power requirements are much higher than other options.

    c. Centrex requires a customer maintain and program the system.

    d. Centrex takes advantage of the built-in redundancy of CO equipment.

2. Why might a company choose PBX over Centrex service?

    a. To use the CO staff for maintenance and programming

    b. Multilocation service

    c. Locked-in pricing

    d. Implement gradual upgrades

3. Centrex service may be combined with other telephone systems, such as PBX and key systems. True or False?

4. Centrex providers supply a customer with all the telephone sets and attendant consoles needed to implement the service. True or False?

5. A company of _____ users will best benefit from Centrex service.

    a. <100

    b. 100-1,000

    c. 1,000-10,000

    d. >10,000

6. Describe how a LEC provides Centrex service to a customer with a single site.

    _____

    _____

    _____

    _____

    _____

Extended Activity

Research the Centrex capabilities provided by your LEC and/or CLEC. Are there any financial benefits of using Centrex over purchasing and using a key system?

Lesson 5—Key Systems

The smallest businesses usually begin with the same sort of single-line telephone installed in most homes. However, as a business grows and adds staff, it needs the flexibility of multiple lines. For a business too big for a single telephone, but much too small for a large-scale office switching system, key systems are the answer.

Key systems are fairly simple on-site telephone systems geared to organizations with fewer than 100 telephones. Like a PBX, they switch calls to and from the public network and within users' premises. However, key systems are simpler than a PBX, reducing the administrative workload for small businesses.

Objectives

At the end of this lesson you will be able to:

*   Describe the components of a key system

*   Describe the type of business that would most benefit from a key system

*   Explain the limitations of key systems

 **Key Point**

Key systems provide multiple telephones access to any of several lines.

Key System Components

The first multiline business telephone system was called the 1A key telephone system. It consisted of a red hold button, four telephone line buttons, and an office intercom button. This system became the workhorse of small businesses, and many of these systems are still installed today.

A key system provides multiple telephone extensions access to a group of single telephone lines. For example, if a small office has six single lines, it can use a key system to access any of those lines from each of its telephones. Each telephone extension would have six buttons (one for each line). To connect a telephone extension to a line, a caller simply presses one of the unlit line buttons. The concept of key systems is illustrated on the Key Telephone System Diagram.

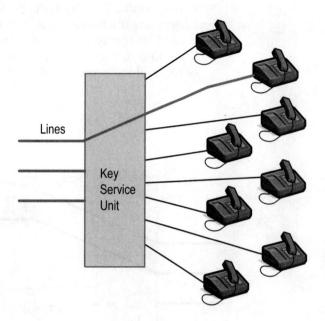

Key Telephone System

All telephone sets in a key system were connected to a central device called a Key Service Unit (KSU), which connected each telephone set to a group of outside business lines. Today, new KSU-less systems offer all the functionality of KSU within each telephone set.

The main point to remember about a key system is that it can support only as many incoming or outgoing telephone calls, or "call paths," as there are lines installed. In other words, if a customer has 100 telephone extensions in an office, but only 40 lines installed, the maximum number of simultaneous calls, coming in or going out, is limited to 40. If the 40 lines are all in use, outgoing callers must wait for a free line, while incoming callers receive a busy signal.

Cordless Key Systems

In many business settings, such as large retail centers or factory floors, cordless key systems provide employees telephone service while allowing them freedom of movement. Wireless transmission is used to connect these mobile extensions to the main business lines, and to each other by means of intercom features. This type of technology is presented on the Cordless Key System Diagram.

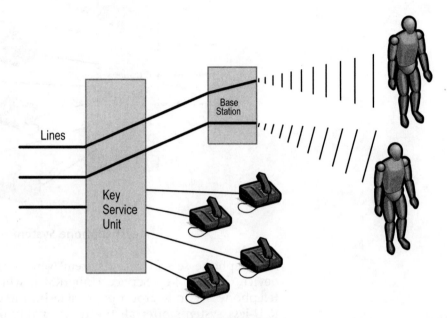

Cordless Key System

The use of wireless telephones inside buildings requires special base stations with antennae located on every floor. There are generally also special outside base stations with antennae for nearby outdoor areas between buildings on a campus. The base stations must be wired with twisted pair to specialized circuit packs within the telephone system cabinet. Specialized wireless telephones associated with key systems and PBXs are high profit margin peripherals. These telephones operate at higher frequencies than home telephones and have specialized features associated with particular key and PBX telephone systems.

On-site wireless telephone systems use a cellular digital switching technology similar to PCS. Calls are transferred between base stations when a user walks out of the range of a particular antenna.

Some mobile telephone units can function both inside and outside of the business campus. They sense when they are out of the range of the base system, and automatically switch calls to a cellular telephone network.

Limitations of Key Systems

A key system provides a cost-effective way for a small business to share a moderate number of telephone lines. However, key systems offer fairly unsophisticated functionality and features. In addition, their main advantage, simplicity, becomes a liability as a business adds telephone lines past a certain point.

As we saw above, each telephone in a key system has a button to access each telephone line, plus a hold button, CO telephone lines, and intercom lines. Therefore, if a business needs 18 business lines, its telephone sets would have at least 20 buttons. By the time a business requires 20 or more lines, each telephone has become quite complex, hard to use, and expensive. Can you imagine an extension telephone with 50 or more buttons on it? Therefore, we can upgrade a true key system only so far before we need to try a different approach.

Hybrid Systems

With the integration of computer technology inside telephone systems, key telephone systems became more and more advanced. Gradually, they began to include features previously found on only full PBX systems. Thus, the term hybrid was used to describe a telephone system that includes features of both a key system and PBX.

A characteristic of a hybrid key system is the grouping of outside trunks into pools, by function or organization. For example, certain trunks are allocated to a particular department.

Electronic Key Telephone Systems (EKTSs) often cross the line into the PBX world, providing switching capabilities, as well as impressive functionality and feature content. EKTS is a key telephone system in which electromechanical relays and switches have been replaced by electronic devices, often in the telephone sets and central cabinet. The inner workings of the central cabinet of an EKTS more resemble a computer than a conventional key system.

This trend, called Computer Telephony Integration (CTI), is widespread. As telephones become more sophisticated, they become more like personal computers (PCs). Currently, telephone set features are melding with those of desktop PCs, especially in IP-based telephone systems. Although telephone sets may never fade away, they continue to become more advanced, featuring video, Internet access, and other visual dialing features.

Activities

1. Explain how a key system provides connectivity to each extension.

   _____

   _____

   _____

   _____

2. Describe the purpose of a KSU.

   _____

   _____

3. How would a key system loose its practicality as a business' communications needs increase?

   _____

   _____

   _____

   _____

4. How does a hybrid key system compare to a standard key system?

   _____

   _____

   _____

   _____

Extended Activities

1. Visit **http://www.nortelnetworks.com** and research the Norstar line of integrated communications systems, which are examples of a hybrid key system.

2. Visit **http://www.vodavi.com** and **http://www.panasonic.com**, and compare their key system features. How do the features these hybrid systems provide compare to a standard key system?

Lesson 6—Buying a Telephone System

A telephone system can be a powerful tool that keeps a business in touch with customers, boosts productivity, and builds revenue. The array of choices has never been wider. Even single-line telephones are packed with features unheard of a decade ago.

Today's sophisticated multiline systems can answer, sort, automatically distribute, and save calls. They can merge voice mail, e-mail, and faxes into a single device. They can connect credit card readers, modems, and cordless and wireless telephones. They can even connect calls to telecommuters working at home, as if they were in the main office.

However, this wealth of choices is bewildering to many business customers. Therefore, a good way to approach buying a telephone system is to systematically divide the task:

- **First plan**—Conduct research to determine the right type of telephone system, and the necessary applications.

- **Then shop**—Find the right vendor at the right price, and arrange for service and support.

Objectives

At the end of this lesson you will be able to:

- Describe the information that should be gathered when planning for a new telephone system

- Describe some of the most important applications included in business telephone systems

- Explain what a Request for Proposal (RFP) is, and why it is necessary

- Discuss the two factors that should determine the selection of a telephone equipment vendor

 **Key Point**

Investing in a telephone system is a vital strategic decision.

Planning: First steps

A thorough planning step may be the single most important factor in buying a telephone system, yet it is often ignored. Before you can choose the right telephone system, you must first define the work the system must do.

Analyze Current Traffic Flow

Begin your planning by preparing a work flow chart that shows how calls are answered, who needs telephones, how often callers are put on hold, and other key aspects of the organization's work flow. Gather statistics on the current volumes and patterns of call traffic flow. Use this information to identify any problems the new system must correct.

Identify Needs and Desires

It is also important to list any desired features the current system does not provide. To identify these needs and wants, it helps to ask questions such as:

- Is the business trying to grow? A new system must be flexible enough to grow as well.

- Do workers need to stay in touch with each other and customers beyond normal business hours?

- What applications are necessary?

- What is the budget, both for the telephone system itself and any add-on applications?

- How much training will be needed to get the most out of the new system?

Gathering good requirements takes time and effort. However, clear requirements make the shopping process easier and more productive.

Choosing a Switching System

Two major types of telephone systems are currently available:

- Key systems are traditionally used by companies with fewer than 50 employees. Key systems are based on the old multi-line telephones that use several buttons, or keys, to access outside lines.

- PBXs handle larger workloads. They are essentially smaller versions of CO switching equipment.

Recently, the distinctions between key and PBX systems have become blurred. Many key systems include features once available only on PBXs, while other systems operate internally as either a key or PBX, depending on the software installed. The term "hybrid" is often used to describe systems that resemble both key and PBX systems.

Digital vs. Analog

Most current telephone systems communicate by means of digital technology. This means the sound is transmitted as bits of data rather than audio waves. Digital transmission has many advantages over analog transmission. Digital signals are less affected by interference and line degradation, thus digital lines have virtually no static or hiss.

However, most businesses make outgoing calls over regular analog lines. This means that a digital telephone system must convert signals back to analog waves whenever a call leaves the office. Because very little sound degradation occurs within the smaller confines of an office, analog systems actually sound about the same as their digital counterparts.

The main reason for buying a digital system is that they tend to be better equipped to connect with accessories such as voice mail or Caller ID. Also, a converged network, one that supports both telephone and data communication, is easier to implement when the telephone switch is already digital.

Sizing a Telephone System

To make sure the new unit is the right size for an organization, it is important to understand the size constraints of different systems.

In key systems, size is usually indicated as a combination of "lines" and "extensions." Lines are the total number of outside lines used by a company. Extensions refer to every inside telephone within the company. For example, a key system might accommodate up to 12 lines and 36 extensions.

In contrast, most PBXs define size in terms of "ports." Ports indicate the maximum number of connections that can be made to the system. This includes outside lines and inside extensions, as well as connections to accessories such as voice mail or automated attendants.

To support the future growth of a business, the ideal telephone system should handle expansions and upgrades in a cost-effective manner. For example, a system should allow the user to replace low-density port cards with those of a higher density before an expansion cabinet or complete system upgrade is required.

Therefore, when evaluating the future costs of owning a system, check which components will need to be purchased or replaced as the organization grows.

System Features

Telephone systems can be equipped with literally hundreds of features for switching calls and directing traffic. However, vendors estimate that 95 percent of these features are never used. Therefore, instead of comparing lists of features, examine how employees actually use the telephone system. Then limit your search to only those features that improve work flow in the office, and that employees are likely to use. This will let you focus on the relevant differences between systems.

Although having the right features is important, it is even more important that those features are easy to use. Most employees devote very little time to learning how to use a telephone system; it is important that the most common functions are extremely simple and intuitive.

Selecting the Telephones

After you have selected the office switching system, it is time to choose the telephone desk sets for inside extensions. In addition to the attendant console, four types of telephones are available for voice-only service:

- Analog, single-line telephones with a standard keypad

- Multiline telephones

- Digital telephones (single and multiline) containing special function buttons in addition to telephone line indicators and selection buttons

- IP telephones, both hard and soft

Multiline vs. Single-Line

Multiline telephones are convenient because they feature a status display of individual lines, and allow several people to answer a common group of lines. Therefore, multiline telephones are appropriate for any area in which users put one call on hold while placing another call, or for switching between two calls. For example, choose multiline telephones for centralized group call coverage, answering positions, purchasing departments, personnel departments, or information service areas. In addition, most business executives need multiline telephones, so they can access both published and private extension numbers.

Most multifunction telephones can have buttons programmed either as functions or individual lines. Lines require status indicators, and the administrator should be aware of this requirement when evaluating programmable, multibutton telephones.

Integrated voice/data communications also require multiple line support. A few single-line digital telephones can support either voice or data operation, but not concurrently.

When deciding which type of telephone to assign to the typical user in your company, consider the tasks that person usually performs. For example, a telephone that supports call waiting, call hold, and conferencing features can usually eliminate the need to have more than one line installed on it.

Trunk Requirements

To keep a system installation on schedule, trunk facilities must be purchased and provisioned by the time system installation begins. In addition to the number of telephones a business estimates using over the next 12 months, you must determine the type of trunks needed to link the business to its local and long distance service provider's CO switching facilities. Based on the organization's business activities, choose some or all of the following trunk categories:

- Public direct outward dialing (DOD) and direct inward dialing (DID)

- Analog private lines ("ear and mouth" [E&M] tie trunks)

- Digital private lines (T1, fractional T1 [FT1], and ISDN-Primary Rate Interface [PRI])

- FX lines

To determine the required number of trunks, you must consider both incoming and outgoing traffic flow for both voice calls and data. The process of analyzing traffic statistics to determine a system's capacity is called traffic engineering, which we discuss in Module 5.

Cost

Many local telephone companies charge one price for telephone lines to a key system, and another for trunks that connect a PBX, even though both have essentially the same functionality. Therefore, check the rates of various types of connectivity before making a buying decision.

Cost is the main reason why businesses usually choose T1 or ISDN-PRI service for their PBX systems; a single T1 line costs less than 24 POTS channels. A T1 provides 24 channels over only two wire pairs, unlike the 24 or more pairs required for the same number of analog channels, and so reduces the required number of PBX physical trunk ports and associated circuit packs. This provides the added benefit of freeing carrier slots to support more station ports, a cost-effective means of expanding the system to support more users.

Features

When comparing T1 to ISDN-PRI, consider the services and features that you need. PRI delivers a wide range of powerful features, but at a correspondingly higher price. If your design simply needs multiple channels, but not the advanced functions of ISDN, then T1 is usually a more economical option. For example, a simple data link, such as an IP connection to an Internet Service Provider, would waste ISDN features such as calling party identification.

Physical Issues

When ordering trunks, you must also consider physical issues, such as the type and number of cables available. If enough copper pairs or fibers are not already in place to provide the service you need, then you must change your design or pay to have additional media installed.

The location of physical media can also affect a design solution. For example, some business parks or office campuses provide centralized cable entrance facilities in one or two buildings. From those building entrances, distribution cables serve the rest of the campus. If you need trunk service to a building that does not have its own service entrance, then you must provision those trunks over the distribution cabling, or install an alternate link such as a microwave system. The device used to connect the transport carrier's outside plant to the distribution cables will vary according to the physical media to be connected and the type of service they carry.

Switch Capacity

Before purchasing trunks, you must also make sure that your PBX hardware and software can accommodate them. Each trunk must connect to a hardware port on the PBX; if the PBX's ports are already used to capacity, the system will not support the projected trunking loads. Consequently, you will need to upgrade the switch hardware before the new trunks can be installed. This could mean upgrading to higher-density port circuit packs, adding an expansion carrier, or replacing the entire PBX.

The switch software may also have a particular trunk capacity; this capacity, like the software itself, may vary according to the features that the software implements. You may also need to increase the number of software licenses or the "right to use."

Number Portability

When an established business changes its location or telecom provider, it naturally wants to retain its existing telephone numbers. Often, Remote Call Forwarding (RCF) can be used to transfer calls from the old serving switch to the new central office. However, RCF may not be available if insufficient trunks link the old and new switches, if the local telecom carrier is running out of NPA or NXX numbers in those areas, or if 911 backward trace cannot be guaranteed.

An RCF solution will often require a different mix of trunks than a non-forwarded approach. Therefore, the availability of number portability must be confirmed as soon as a business decides to move or change phone companies. If portability is not available, the business must budget time and money to change its marketing literature, and notify its customers and vendors.

Determining Required Applications

Companies of all sizes need more than simple telephone switching to conduct effective, productive, customer-focused communications. Now, businesses need true communications systems that include advanced software applications and add-on equipment to customize business communications.

The following applications are quickly becoming required features of advanced telephony systems:

- Automatic Call Distributor (ACD) systems
- Messaging—Automated attendants, voice mail, and Interactive Voice Response (IVR)
- Wireless systems
- CTI
- Internet Protocol (IP) Telephony

When evaluating the applications offered with various systems, take advantage of comparison shopping information available in magazines and on the Internet. For example, look for reviews in consumer and telecommunications publications, or articles that report the results of equipment "test drives."

ACD Systems

Automatic Call Distributor (ACD) systems route a high volume of incoming calls to a pool of call-answering agents. ACDs are generally used in call centers that process many incoming calls. When an ACD is added to a communications system, it can provide important operational efficiencies. This is especially true for mail-order businesses or other service companies with high-volume calling.

A typical ACD system provides call routing, queuing (placing callers on hold, in order), and call statistics to help you understand who's calling, when, and how long they stay on the line. Advanced ACDs can also use Caller ID to pull customer records from a computer database when a call comes in.

Many ACD systems are connected with voice mail or IVR systems to reduce agent workloads and help redirect calls. Recorded announcements may also be used to guide callers to the proper extensions. Some systems give the caller a choice of call destinations based on menu selections.

ACDs can perform call routing based on a variety of factors unique to your business, such as the skill level of individual agents, location a call originated from, time of day it came in, how long the caller has been in the queue, and availability of other agents to handle the call.

Messaging Systems

Messaging applications are widely used in businesses of all sizes. Several types of messaging products are designed to work with a basic communications system. These include automated attendant systems, voice mail systems, IVR systems, and speech recognition systems. Many manufacturers now provide support for multimedia messaging, such as combined voice mail, fax, e-mail, and video.

Automated attendants are especially useful for reducing the incoming traffic load for operators, receptionists, and call center agents. Automated attendant systems generally offer callers a voice menu for automatically routing a call to the desired destination.

Messaging systems support a wide range of services, including automated routing, voice mail, and even informational services, such as product announcements or directions to your facility. Most messaging systems can be accessed from remote locations, thus traveling employees are always in touch.

Some systems also offer message broadcasting, which delivers the same voice mail message to multiple recipients. For particularly urgent messages, paging services can alert employees to check their voice mail. Multiple messaging systems can be linked together to support high-volume or multiple-location businesses.

Wireless Systems

Wireless, or "mobility," systems are now offered by most major vendors. Unlike cellular phones, these systems work in connection with a PBX or key system, and replace a hard-wired extension line with a radio link. Mobility systems usually include a controller, antennas for radio transmission and reception, and portable handsets/headsets. They have few integrated features, and rely instead on the main communications system for functions such as call hold, call transfer, or voice mail.

Wireless telephones are beneficial in locations where wired telephones are not cost-effective, such as outdoor campus areas, or where employees are naturally mobile, such as large warehouses. Industries that use wireless systems include health care, schools, construction firms, car dealers, restaurants, realty offices, and other service companies.

CTI

Computer Telephony Integration (CTI) is the merging of telephone and computer systems to expand the power of a call center and improve the speed and efficiency of customer service. In a typical call center environment, CTI enables calls to be automatically assigned to the most effective, available agent or system. It also prepares the agent with all information necessary to handle a specific caller or type of call. When a customer calls in, the account information automatically pops up on the agent's computer screen, streamlining the process of handling customer requests.

CTI also lets businesses personalize options for callers, and eases the process of transferring callers and their account information between agents and departments.

IP Telephony

More and more today, businesses want to consolidate different types of communication traffic, such as voice calls, data transmission, and video conferencing, onto a single network infrastructure. This can simplify the communications process (fewer lines and network providers to manage) and cut the costs of calls.

IP Telephony is helping to make this possible, by allowing standard public-network calls to be carried over packet networks like the Internet. Through IP Telephony, businesses can save significantly on both voice calls and fax services. Newer software now allows remote and traveling workers to take advantage of IP Telephony from either desktop or laptop computers.

Finding the Right Vendor

Virtually all telephone systems require the assistance of a vendor for programming and installation. As a result, finding a good vendor can be the most important part of the purchase, because any telephone system must be properly installed for optimal performance.

Qualifying the Vendors

To qualify prospective companies and sales representatives, a customer should start by gathering personal recommendations. Ask business associates to name the telecommunications equipment company they chose, and explain the reason for that choice. Check the Better Business Bureau and get references.

Submitting an RFP

When you have identified three to five vendors and systems that meet your needs, you are ready to ask the vendors for written proposals. A proposal is a written statement of the vendor's installation plan. It includes details of system capabilities and costs, installation time frames, and service arrangements.

Proposals are obviously much more useful if they follow a similar format and content that allows easy comparisons between vendors. To make sure this happens, a customer should write a request for proposal (RFP). The RFP tells each vendor the type of proposal a customer is interested in receiving, and includes the information a vendor must know to make appropriate recommendations.

RFPs vary greatly. Some RFPs are very specific, and ask vendors to bid on exact lists of equipment and specific physical configurations. Others define the mandatory, important, or desirable features, and ask vendors to propose their own solutions. In general, an RFP for a telephone system should include the following elements:

- **Overview and instructions**—Explanation of the rules of the selection process. An RFP should clearly state whether the customer is interested in a vendor's suggestion of an alternate design; if you welcome other designs, include guidelines that say how you want those designs to be submitted.

- **Schedule of deadlines and presentation dates**—Dates by which proposals must be received to be considered.

- **Definition of evaluation criteria**—Factors that will determine the customer's choice. For example, will the winning bid offer the lowest price or the best solution?

- **Objectives of the project**—Goals the new system must achieve.

- **Overview of current technology**—Description of the current system and its performance.

- **Overview of proposed technology**—Description of the desired system, and specification of required features. If necessary, requirements can be sorted into mandatory, desirable, and wish list items.

- **Required information and format**—For example, many RFPs require a vendor profile, financial statement, description of experience with similar installations, and three client references.

Selecting the Vendor

Based on the vendors' proposals, invite the best candidates to make an oral presentation and demonstration. Finally, award the contract to your best choice. However, do not forget about the other candidates, and save their proposals; they could be helpful if problems arise.

Because a telephone system is a mission-critical element of any business, a low-priced system is a bad choice if the vendor provides poor installation or support. Therefore, the final choice of a vendor should be based on both price and quality.

Pricing

While the smallest systems may cost a few thousand dollars to install, the price tag for more complex models can quickly climb to tens of thousands of dollars. Telephone system prices vary based on four factors:

- **Central cabinet**—The central cabinet contains the switching system that controls and oversees the entire telephone system. Prices differ between systems and increase as cards and accessories are added. A small central cabinet can cost as little as $3,000, with the price increasing considerably for larger systems.

- **Telephone desk sets**—Most systems can be equipped with several types of telephones. The least expensive sets may cost less than $100, but can make it very difficult to access features. Some "executive telephones" sell for many times the standard price. Although these telephones can make it slightly easier to use the system, they are more often a significant source of profit for the vendor.

- **Wiring and installation**—Installation costs depend on the physical conditions at the site. For example, it is usually inexpensive to install wires in an unfinished building. It is much more costly to install wiring through already finished walls. To avoid rewiring in the future, specify that plenty of wiring be installed when the system is first purchased. A good rule of thumb is to request at least twice the wiring you currently need. Although this adds to the cost of installation, the additional cost is only a fraction of the cost of adding wires later.

- **Everything else**—This includes training, programming, service, and future modifications. Because pricing is usually based on the time these tasks require, this can be the most flexible portion of a bid. It is often good to compare the estimated hours for these tasks with the price tag for each service.

Vendor Experience

The most important consideration in choosing a vendor is the number of installations the vendor has completed using your desired system. When references indicate the vendor is committed to a product line, you can be assured of a long-term source for service. A vendor who has installed many of the same systems will be much more familiar with their typical problems.

Customers should also ask each vendor to describe its installation experience, and provide references to support its claims. A vendor's references should include installations completed in the past year, and describe the sizes of the companies involved and the options or features that were added. Most important, a customer should take the time to call each vendor's references, to understand the specific nature of each project and discuss the vendor's strengths and weaknesses.

Shopping for Service and Support

Selecting the right post-warranty service plan is key to the success of a new telecommunications solution. When evaluating the options available for post-sale service and support, customers should get answers to the following questions:

System Setup

- Who will do the installation? Is it being subcontracted out? How long will it take?

- Who will provide live training? How many hours are included in the quoted price, and how long is training estimated to take?

- What other types of training (compact disc-read only memory [CD ROM], video, etc.) are offered? How much do they cost?

Warranty
- What kind of coverage does the warranty provide? What does it specifically not cover?

- What kind of coverage is provided in the post-warranty service plan?

- Can the total package include an ongoing service and support plan for the period after the warranty expiration?

- Does the system include appropriate protection against power surges or lightning? In the event a system and/or application is damaged by a power surge or lightning, will the customer be reimbursed for all damage, up to and including a full replacement?

Service Level Commitment
- What is the on-site service commitment? Specifically, how soon will a technician be on site for major alarms? Minor alarms? How are major and minor alarms defined?

- What is the commitment for out-of-hours support? When the business requests an on-site service call after hours, how soon will a technician arrive?

- How fast are replacement parts available? Same day? Overnight? Are there extra charges for this?

- Where is the nearest parts distribution facility? Are parts available locally, or must they be shipped from another city or region?

- Can the system be programmed or administered remotely? If so, how is this service billed? What is the response time? What hours is this service available?

Technical Support
- Does the vendor offer a technical support Help line? When is the service available, and how experienced are the support agents? Is there a limit to the number of free support calls? If so, how much does it cost to exceed the limit? If a customer does not have a service plan, what is the charge for each call?

- Do customers have full access to a Web-based Customer Service Support site? Does it provide important services such as ordering replacement equipment, creating a service request, checking on the status of a request, downloading software, and providing electronic product information and technical bulletin boards?

System Security

- Does the service plan include screening for potential toll fraud? Does it include active intervention if toll fraud is occurring? If so, who will do that, and what training has that person had?

- What is the charge for toll fraud intervention service without a service plan?

Preventive Maintenance

- Does the service plan include all labor charges for proactively scheduled preventive maintenance?

- Is there an extra charge for performing this work outside of business hours?

Activity

You are assigned to analyze and choose a new telephone system for your organization. Your system will be for 100 telephone users, averaging 10 calls per day with an average of 2 minutes per call. The business is growing rapidly and plans to double over the next two years. The phone system should accommodate at least this two year growth period.

The system is to be available 24 hours per day, 7 days per week. Each employee needs voice mail, do not disturb, and paging features.

Compare at least two telephone systems and make a recommendation.

_____

_____

_____

_____

_____

_____

Extended Activity

1. Based on the problem above, determine the following for at least one of the systems recommended:

   a. Price per phone user for the system.

   _____

   b. Warranty

   _____

   c. Installation costs

   _____

   d. Technical support methodology

   _____

Summary

In this module, we reviewed the structure of the public telephone network, described some of the equipment required to connect business customers to the network, and discussed a systematic process for purchasing business telephone switching systems.

In Lessons 1 through 3, we saw that the telecommunications system is made up of the following elements:

- Telephone COs, owned by various business entities, and organized in a hierarchy of connections

- Telephone numbering system that provides a unique address to each customer

- Transmission paths (lines and trunks) that transmit signals between customers and COs

We learned that the telephone system is organized into a hierarchy of switching offices, and saw how calls are passed between competing local carriers, and between local carriers, resellers, and IXCs. Each CO provides service to individual subscribers in its geographic area. The ILEC owns most of these offices; however, CLECs may offer alternative services by building their own offices and connecting them to the ILEC's network.

In Lesson 3, we discussed the various types of transmission connections provided by the telephone network. We learned that lines serve individual telephones, while trunks connect two switches, such as CO switches or business PBX systems. Furthermore, we saw that businesses can order trunks that support different types of transmission (analog or digital) and information-carrying capacities.

Lessons 4 and 5 described two types of switching systems that route telephone calls to and from a customer's business premises. A key system is one of the simplest business telephone switching devices. If a business uses multiple single telephone lines, a key system allows multiple inside telephones to connect to any of those lines.

Larger businesses can enjoy the full range of telephone switching features, without purchasing their own equipment, by using the Centrex service. In this approach, each business is connected to the CO switch by a separate telephone line. The CO switch then connects calls to and from each of those telephones, and provides services such as three-way calling.

The third option for business switching, a PBX, is such a popular approach that this course devotes all of Module 2 to discussing PBX concepts, features, and configuration.

In the meantime, Lesson 6 summarized a process for choosing a business telephone system. We saw that a systematic approach, based on objective requirements and written proposals, is the best way to ensure that an organization gets a high-quality, cost-effective system that meets its current and future needs.

Module 1 Quiz

1. A class 5 office connects individual subscribers to the telephone network and is also referred to as a(n) _____.

    a. End Office

    b. Central Office

    c. Local Office

    d. All of the above

2. Central Offices are connected to which of the following?

    a. Class 4 Offices

    b. Class 5 Offices

    c. Individual subscribers

    d. All of the above

3. Which of the following is closest to the telephone subscriber?

    a. Class 1 Office

    b. Another subscriber

    c. Class 4 Office

    d. Class 5 Office

4. The switch within a CO that sends traffic to another LATA is referred to as a _____.

    a. POP

    b. Router

    c. Gateway

    d. DACS

5. Each wire of a local loop is sometimes referred to as _____.

    a. Plug and Play

    b. Tip and Ring

    c. Plug and Socket

    d. Ring and Plug

6. When two tones are used to signal switching equipment it is called:

   a. Tip and Ring

   b. Step-by-step

   c. DTMF

   d. Out-of-band signaling

7. What is the switch that is used to connect two COs?

   a. Routing switch

   b. Interconnect

   c. Layer 5 switch

   d. Tandem switch

8. When signaling takes place over a separate path than the path the analog voice signal takes, it is referred to as _____.

   a. Out-of-band signaling

   b. In-band signaling

   c. SS7

   d. Dial tone signaling

9. What is the digital network used to create the route a call will take?

   a. Out-of-band signaling

   b. In-band signaling

   c. SS7

   d. Dial tone signaling

10. Overlay codes are used because of_____.

   a. Too many COs

   b. Too many Tandem Offices

   c. Too many telephone numbers used

   d. Too many subscriber services

11. The concept of keeping the same telephone number when switching from a LEC to a CLEC is called _____.

    a. Local Number Portability

    b. Number switching

    c. CO Number Plan

    d. CO Number Portability

12. The physical connection that links the customer premises to the telephone company CO switch is referred to as a _____.

    a. Local loop

    b. Trunk

    c. Channel

    d. Digital Circuit

13. Line can mean which of the following?

    a. Physical connection to an individual telephone

    b. Transmission path between a CO and a customer

    c. Transmission path between a PBX port and an individual telephone

    d. All of the above

14. The connection between switching systems is referred to as a _____.

    a. Local loop

    b. Trunk

    c. Channel

    d. Digital Circuit

15. What are the two basic qualities of an analog signal?

    a. Frequency and phase

    b. Phase and amplitude

    c. Frequency and amplitude

    d. Amplitude and electromagnetism

16. The pitch of a sound is determined by its _____.

    a. Amplitude

    b. Phase

    c. Distance to the telephone set

    d. Frequency

17. The loudness of a sound is determined by its _____.

    a. Amplitude

    b. Phase

    c. Transmission type

    d. Frequency

18. What is the sampling technique used to convert analog voice signals to digital signals?

    a. Pulse Code Modulation (PCM)

    b. Amplitude conversion

    c. Phase conversation

    d. Frequency modulation

19. Multiple digital signals share the same physical media through a process known as _____.

    a. Modulation

    b. Multiplexing

    c. Conversion

    d. Distortion

20. A dedicated CO circuit that links two points is referred to as a _____.

    a. Trunk

    b. POTS line

    c. Tie Line

    d. FX line

21. A line that goes from a customer's site through a CO to a remote CO in another part of the country is known as a _____.

    a. Trunk

    b. POTS line

    c. Tie Line

    d. FX line

22. What is a telephone line that is dedicated between two telephones?

    a. Trunk

    b. POTS line

    c. Ring down

    d. FX line

23. When a CO switch provides a business telephone switching services, it is referred to as _____.

    a. DACs

    b. NAR

    c. Centrex

    d. CO provisioning

24. What is one of the advantages of a Centrex service?

    a. The business does not have to take care of maintenance and programming of the telephone system

    b. The business saves on physical space

    c. The business does not have to worry about equipment reliability

    d. All of the above

25. What is one of the limitations of Centrex?

    a. The business locations must be located within a single LEC area

    b. Ease of upgrades

    c. Vendor support

    d. Compatibility

26. In a Centrex system, the individual lines are also known as
_____.

   a. DACS

   b. NARs

   c. CLECs

   d. Tie lines

27. What is one advantage of Centrex over a PBX system?

   a. Ease of billing

   b. On-site switching

   c. Service to telephone set

   d. Rental (no property tax)

28. What is the device that provides multiple telephone extensions access to a group of single telephone lines?

   a. PBX

   b. FX

   c. Key system

   d. Cordless system

29. If a customer has 50 telephone extensions and 20 lines installed, what is the maximum number of simultaneous calls that can be made with a key system?

   a. 100

   b. 50

   c. 20

   d. 70

30. A hybrid telephone system combines the features of which of the following two systems?

   a. PBX

   b. Cordless

   c. Key system

   d. Multiplexer

31. What is a key system where electromechanical relays and switches are replaced by electronic devices?

    a. PBX

    b. Cordless systems

    c. Multiplexers

    d. EKTS

32. When PBX and computer systems are connected, the result is referred to as _____.

    a. IP Telephony

    b. Competitive LEC

    c. CTI

    d. Centrex

33. A voice call routed over the Internet is referred to as _____.

    a. IP Telephony

    b. Competitive LEC

    c. CTI

    d. Centrex

34. What are the two major types of phone systems?

    a. Key and PBX

    b. Digital and Analog

    c. PBX and digital

    d. PBX and ACD

35. What is the first step in buying a new phone system?

    a. Choosing the switching system

    b. Analyzing the telephone traffic patterns

    c. Deciding between analog and digital

    d. Determining the trunk requirements

Module 2
PBX and ACD Systems

This module focuses on the most important workhorse of any business: the private branch exchange (PBX) telephone switching system. Lessons 1 and 2 explain how a PBX operates, and describe the features and functions common to most private business telephone switches. The Automatic Call Distributor (ACD), once an optional add-on to a PBX, has become essential to businesses and expected by many customers. Lesson 3 introduces the basic concepts and terms that describe ACD functions, and explains the relative advantages and disadvantages of the most common call routing methods.

Lessons 3 and 4 concentrate on one type of PBX, the DEFINITY Enterprise Communications Server (ECS) developed by Avaya Technologies. We will reinforce basic PBX concepts by learning about the logical architecture and physical installation of a DEFINITY system.

Voice over IP (VoIP) and IP telephony are developing beyond the initial stage of buzzword craze into reality. Today, businesses are realizing the benefits and cost savings that are possible using IP communication equipment and converged network environments. Lesson 5 reviews the IP telephony hardware and software components. It first looks at the software that provides call processing functions. It then reviews media servers, media gateways, the IP office product and telephones that are supported in this new environment.

Seamless communication between multiple locations is a vital requirement for many businesses. Lesson 7 examines common approaches to implementing private communication networks that parallel the public-switched telephone system. We will see how a combination of telephone switching and dedicated lines can create a wide area voice and data network that can treat most intraoffice calls as "local."

Lessons

1. PBX Fundamentals

2. PBX Features and Functions

3. PBX Configurations

4. DEFINITY Server Hardware

5. IP Telephony

6. PBX Networking

7. Basics of Automatic Call Distributor

Terms

802.1x—IEEE 802.1x is a recently approved IEEE standard for port-based access control. It is used to control access to a network access device (switch, access point, etc.).

802.3af—IEEE 802.3af is a standard proposed by the Institute of Electrical and Electronic Engineers (IEEE) for powering Ethernet devices over twisted pair cabling. IEEE 802.3af is a legacy Ethernet-compatible, internationally standard power distribution technique.

Adjunct Processor—An adjunct processor is an external computer that controls a PBX. For example, a CTI server is an adjunct processor. In the Avaya DEFINITY PBX, an API called ASAI makes it simpler to write CTI link applications that allow a CTI server to control a DEFINITY PBX.

Adjunct-Switch Application Interface (ASAI)—ASAI is a Avaya- specific API for its DEFINITY PBX switch. ASAI provides commands and messages that enable features such as event notification and call control. CTI link applications that use ASAI can more easily access these services on a DEFINITY system.

Agent—A person or automated device that serves incoming callers is referred to as an agent.

Agent State—Agent state refers to the current availability status of an agent. The term also represents a user's ability to change an agent's availability within the system.

Americans With Disabilities Act (ADA)—The ADA is a federal law that guarantees equal opportunity for individuals with disabilities in public and private sector services and employment. Generally, the ADA bans discrimination on the basis of disability, and requires employers to make "reasonable accommodation" to allow the employment of people with disabilities who are otherwise qualified to perform the essential functions of a job.

Announcement—An announcement is a prerecorded message delivered to a caller in queue requesting the caller to remain online, prompting the caller for information, or directing the caller to another destination. Announcements can be scheduled to occur in a particular order, or repeat periodically.

Asynchronous Transfer Mode (ATM)—ATM is a connection-oriented cell relay technology based on small (53-byte) cells. An ATM network consists of ATM switches that form multiple virtual circuits to carry groups of cells from source to destination. ATM can provide high-speed transport services for audio, data, and video.

Auto attendant—An auto attendant is a device, or PBX function, that answers incoming calls, plays a greeting and instructions, then allows callers to directly dial an extension number, or the attendant. This feature can be used to relieve receptionists during peak calling periods and breaks; however, many companies use automated attendants instead of receptionists.

Automatic Call Distributor (ACD)—An ACD is a programmable system that controls how inbound calls are received, held, delayed, treated, and distributed to call center agents.

Automatic Number Identification (ANI)—ANI is a trunk-based feature that passes the number of a calling party to the telephone receiving the call. This information can be sent in several formats, both analog and digital. Some circuits send the information as a series of tones between the first and second ring. Others send the information as a series of DTMF tones after the call is answered. Still other digital circuits (such as ISDN) send this information on a separate digital channel using SS7. Although similar to Caller ID, ANI cannot be blocked by the calling party.

Avaya Access Security Gateway (ASG)—ASG is a suite of products designed by Avaya to secure access to IP and communication systems.

Avaya Communications Manager (ACM)—ACM is the next generation of Avaya call processing software. ACM runs on legacy DEFINITY servers, as well as on Avaya's new line of media servers. ACM is a component of Avaya's Enterprise Class Internet Protocol Solutions (ECLIPS) product line.

Best Service Routing (BSR)—BSR is an ACD feature that allows the ACD to determine which split or skills can provide the best possible service to the caller. BSR can be used in a single site or across multiple sites, where it works in conjunction with LAI to send the call to the best qualified resource (either local or remote).

British Thermal Unit (BTU)—A BTU a standard measure of thermal energy. The heat output of electronic equipment is specified in BTUs per hour.

Bus—A bus connects the central processor of a PC with the video controller, disk controller, hard drives, and memory. Internal buses are buses such as AT, ISA, EISA, and MCA that are internal to a PC but not "local buses."

BX.25 Protocol—BX.25 is a Lucent/Avaya proprietary variation on the X.25 packet-switching protocol. BX.25 is used in Lucent Distributed Communications System and Digital Communications Interface Unit links.

Call Center—A call center provides a centralized location where a group of agents or company representatives communicate with customers by means of incoming or outgoing calls.

Call Coverage—Call Coverage provides automatic call redirection, based on specified criteria, to alternate answering positions in a Call Coverage path. A coverage path can include a telephone, an attendant group, an Automatic Call Distribution (ACD) hunt group, a voice messaging system, or a Coverage Answer Group established to answer redirected calls. In addition to redirecting a call to a local answering position, Call Coverage can redirect calls based on time-of-day, redirect calls to a remote location, and allow users to change back and forth between two lead-coverage paths from either an on- or off-site location.

Call Management System (CMS)—CMS is an adjunct (basic software package or optional enhanced software package) that collects call data from a switch resident ACD. CMS provides call management performance recording and reporting. It can also be used to perform some ACD administration. CMS allows users to determine how well their customers are being served and how efficient their call management operation is.

Call Prompting—Call Prompting is a call management method that uses specialized call vector commands to provide flexible handling of incoming calls based on information collected from a caller. One example is when a caller receives an announcement and is then prompted to select (by means of dialed number selection) a department or option listed in the announcement.

Call Treatment—Call treatment refers to a call center's automatic call-handling procedure. Each call treatment is programmed into the ACD, using a simple scripting language. Most call treatments include elements such as waiting time, announcements, on-hold music, and telephone system instructions. Separate treatments are normally programmed for office hours, after-hours, or special events.

Call Vectoring—Call Vectoring is an optional software package that allows processing of incoming calls according to a programmed set of commands. Call Vectoring provides a flexible service allowing direct calls to specific and/or unique call treatments.

Center Stage Switch (CSS)—A CSS is the central interface between the PPN and EPNs. The CSS serves as a hub for WAN carriers that connect multiple locations to the same central switch.

Centralized Attendant Service (CAS)—CAS is a system feature used when more than one switch is employed. CAS is an attendant or group of attendants that handles the calls for all switches in a particular network.

Cladding—Cladding is the clear plastic or glass layer that encloses the light-transmitting core of a fiber optic cable. The cladding has a lower refractive index than the core, and reflects the light signal back into the core as the light propagates down the fiber.

Class of Service (CoS)—CoS is used to manage traffic in a network by grouping similar types of traffic (such as e-mail, video, voice) together and treating each type as a class with its own level of service priority.

Commutator—An electronic device that converts alternating current to direct current is referred to as a commutator.

Compact Modular Cabinet (CMC1) Media Gateway—A CMCI is an Avaya media gateway that consists of a floor or wall mount compact cabinet containing control and interface circuit packs. A CMC1 can support the DEFINITY server CSI and the S8100/8500/8700 media servers.

Converged Network—A converged network is a network that supports both data and voice communication, or data and multimedia.

Core—The core is the innermost layer of a fiber optic cable, made of clear glass or plastic. The core carries light signals down the fiber.

Coverage Answer Group—A coverage answer group is a group of up to eight voice terminals that ring simultaneously when a call is redirected to it by Call Coverage. Any one of the group can answer the call.

Coverage Call—A converge call is a call that is automatically redirected from the called party's extension to an alternate answering position when certain coverage criteria are met.

Coverage Path—The order in which calls are redirected to alternate answering positions is referred to as the coverage path.

DEFINITY Enterprise Communications Server (ECS)—DEFINITY ECS is a digital switch that processes both voice and data traffic.

Denial of Service (DoS) Attack—A DoS attack is an attack on a network that is designed to cause slowing or collapse by flooding the network with useless traffic. These attacks usually exploit limitations in the TCP/IP protocols.

Dialed Number Identification Service (DNIS)—DNIS is a feature of toll-free service (800/888/877) that sends the dialed digits to the called destination. This can be used with a display voice terminal to indicate the type of call to an agent. For example, the destination number can classify a call or caller, depending on the product or service the destination number is associated with.

Digital Announcement Unit (DAU)—A digital announcement unit (DAU) plays a variety of recorded messages to callers on hold, according to the call treatment script. (When customers are waiting in a queue to talk to an agent, they will wait more patiently if they are politely encouraged to do so.) For example, a DAU typically plays an initial greeting that asks customer to stay on the line and wait for the first available agent. Subsequent messages can reassure callers that they will soon be served, or invite them to exit the queue by leaving voice mail messages for agents.

Direct Inward Dialing (DID)—DID is a process by which a PBX routes calls directly to a particular extension (identified by the last four digits). Incoming trunks must be specifically configured to support DID.

Direct Inward System Access (DISA)—DISA is a PBX feature that allows an outside caller to dial directly into the PBX system, then remotely access the system's features and facilities. DISA is typically used to allow employees to make long-distance calls from home or any remote area, using the company's less expensive long-distance service. To use DISA, an employee calls a special access number (usually toll-free), and then enters a short password code.

Distributed Communications System (DCS)—DCS is a Lucent/Avaya proprietary switch and messaging system networking scheme in which up to 20 devices may be networked together passing digital signaling information over these links.

Distributed Communications Systems + (DCS+)—DCS+ is an Avaya proprietary signaling protocol that enhances DCS by passing control information over public networks, such as ISDN-PRI. DCS+ signaling travels between DCS nodes over the ISDN-PRI D-channel.

Dual Tone Multifrequency (DTMF)—DTMF is the "touch tone®" system of signaling telephone numbers. A telephone set has a pad of 12 buttons. Each button transmits a combination of two pure tones: one high-frequency, and one low-frequency.

Dynamic Random Access Memory (DRAM)—DRAM is the memory of a computer that can be read from or written to by computer hardware components such as a CPU. DRAM is volatile memory; data is erased from DRAM when a computer loses power or shuts down.

Electronic Tandem Network—A private wide-area telephone network that uses leased lines to link multiple PBX systems is referred to as an electronic tandem network. Each PBX can serve as a tandem switch, routing calls to any other PBX. The tandem network functions as one unified telephone system.

Encryption—Encryption is the process of scrambling data by changing it in a series of logical steps, called an encryption algorithm. To increase security, an encryption algorithm uses a numerical pattern, or "key," to guide the scrambling process. Different algorithms and keys will each produce data scrambled, or encrypted, in different patterns.

Expansion Port Network (EPN)—An EPN contains additional ports that increase the number of connections to trunks and lines. Unlike a PPN, an EPN does not contain an SPE.

Expert Agent Selection (EAS)—EAS is an optional feature available with Generic 3 and Generic 2.2 that uses Call Vectoring and ACD in the switch to route incoming calls to the correct agent, on the first try, based on skills.

Extension—Voice terminals connected to a PBX/switch by means of telephone lines are referred to as extensions. The term also defines the three-, four-, or five-digit numbers used to identify the voice terminal to the PBX/switch software for call routing purposes.

Facility Restriction Level (FRL)—An FRL is assigned to a subscriber set to control the calling privileges allowed. FRLs work with call permissions and route patterns to control call placement.

Firewall—A firewall is a controlled access point between sections of the same network, designed to confine problems to one section. A firewall is also a controlled access between a private network and a public network (such as the Internet), usually implemented with a router and special firewall software.

Flash Read Only Memory (ROM)—Flash ROM is nonvolatile computer memory that can be erased or reprogrammed; however, it retains its data when power is lost or a computer shuts down.

G.711—G.711 is one of a series of ITU-T voice digitizing algorithms. G.711 transfers digitized audio at 48, 56, and 64 Kbps.

G.723—G.723 is one of a series of ITU-T voice digitizing algorithms. G.723 transfers digitized audio at 5.3 or 6.3 Kbps.

G.729—G.729 is one of a series of ITU-T voice digitizing algorithms. G.729 transfers digitized audio at 8 Kbps.

Grade of Service (GoS)—GOS is the probability that an incoming call will be blocked during the busy hour. For example, a grade of service of 0.001 means that, on average, one call in one thousand will be blocked (receive a busy signal).

H.323—H.323 is the ITU-T recommended set of standards and protocols used to support multimedia conferencing on packet-based, best effort networks, such as the Internet and IP LANs. Included in the recommendation are several protocols that control call setup and teardown and voice and video signal conversion and compression.

Holdover Time—Holdover time is the time the batteries power equipment in the absence of line power.

Hunt Group—A group of trunks/agents selected to work together to provide specific routing of special purpose calls is referred to as a hunt group.

Induction Heater—An induction heater is a device that generates heat by placing a heat-conducting material in a fluctuating electromagnetic field. The changing field induces a circulating flow of electrical energy in the conductor, which produces heat.

Integrated Services Digital Network (ISDN)—ISDN is a digital multiplexing technology that can transmit voice, data, and other forms of communication simultaneously over a single local loop. ISDN-BRI provides two "bearer" channels (B channels) of 64 Kbps each, plus one control channel (D channel) of 16 Kbps. ISDN-PRI offers 23 B channels of 64 Kbps each, plus one D channel of 64 Kbps.

Interactive Voice Response (IVR)—IVR is an interface technology that allows outside callers to control a computer application and input information using their telephone keypads. All IVRs can speak back the results of the computer application, and some can also be programmed to fax back the results.

Interflow—Interflow refers to the redirection of a call to a destination outside the local switch network (different switch system). Interflow is used when a split's/skill's queue is heavily loaded or when a call arrives after normal work hours.

Internet Protocol Security (IPSec)—IPSec was proposed by the IETF as a set of protocols and procedures for securing IP packet traffic on publicly accessible IP networks. IPSec supports several cipher algorithms, and enables both payload authentication and encryption.

Intraflow—Redirection of a call to a destination within the local switch network (same switch system) is referred to as intraflow. Intraflow is used when a split's/skill's queue is heavily loaded or a call arrives after normal work hours.

ITU-T V.92—ITU-T V.92 is dial-up modem specification that features quick connect, modem-on-hold and PCM Upstream. The quick connect feature shortens the modem handshaking portion of the call's setup procedures. Modem-on-hold allows a caller to put a modem call on-hold to pick up an inbound voice call (used in conjunction with call waiting on the modem telephone line). PCM upstream promises upstream speeds of up to 48Kbps.

Leave Word Calling—Leave word calling is a system feature that allows messages to be stored for any ACD split/skill, and allows for retrieval by a covering user of that split/skill or a system-wide message retriever.

Linux—Linux is an open-source (freeware) version of UNIX created as a class project by a Finnish graduate student named Linux Torvalds.

Local Survivable Processor (LSP)—An LSP is an Avaya media server co-located with a remote G700 media gateway that operates as a hot spare in the remote location in the case that the media gateway loses communications with the primary call processor. In the instance of a communication failure between the media gateway and the primary call processor, the LSP takes over call processing for the local site. If the link between the primary and remote site caused the original failover to the LSP, the remote site will not be able to place calls to the primary site over the failed link until the link is restored to service.

Look Ahead Interflow (LAI)—LAI is an ACD feature that allows a busy contact center to move some or all calls to another ACD better able to service them. Vectors define call conditions that can trigger the ACD to send the call to another site, and the receiving site can choose to accept or reject the call. The call remains in queue at the sending location while it determines whether to forward the call on or hold it for the next available agent.

Multicarrier Cabinet (MCC1) media gateway—An MCC1 is a 70 inch cabinet that houses up to five carriers, a fan unit, and a power distribution unit. Each carrier contains control, interface, port, and/or service circuit packs. An MCC1 can be a PN, EPN, or an auxiliary cabinet.

National Electrical Code (NEC)—The NEC is a set of safety standards and rules for the design and installation of electrical circuits, including network and telephone cabling. The NEC was developed by a committee of ANSI, and has been adopted as law by many states and cities. Versions of the NEC are dated by year; each local area may require compliance with a different version of the NEC.

Network Control/Packet Interface (NetPkt)—The NetPkt board connects the process and port circuit packs to the TDM bus and provides an interface to the processor for D-channel signaling over the packet bus. The NetPkt board replaces both the NETCON and PACCON boards in newer systems.

Network Controller (NETCON)—The NETCON board is used in older systems; it connects the process and port circuit packs to the TDM bus.

Night Service—Night service is used when a call arrives after normal work hours. The call can be redirected to another destination, such as another split/skill, an extension, the attendant, an announcement with forced disconnect, or a message center. Night Service can take one of three forms:

• Hunt Group (Split/Skill) Night Service

• Trunk Group Night Service

• System Night Service

Node—A switching or control point for a network is referred to as a node. Nodes are either tandem (they receive signals and pass them on) or terminal (they originate or terminate a transmission path).

Nonvolatile Memory—Nonvolatile memory is computer memory that retains its data when power is lost or a computer is shut down. See Flash ROM.

Occupational Safety and Health Administration (OSHA)—OSHA is the federal agency that establishes and enforces measurable standards for workplace safety.

Outcalling—Outcalling is a voice mail feature that dials an outside number (such as a pager), that alerts a traveling user to new voice mail.

Packet Bus—The packet bus runs internally throughout each PN, and terminates on each end. It is an 18-bit parallel bus that carries logical links and control messages from the SPE, through port circuits, to endpoints, such as terminals and adjuncts. The packet bus carries X.25 links, SNIs, and remote management terminal traffic.

Packet Controller (PACCON)—The PACCON board is used in older systems; it provides an interface to the processor for D-channel signaling over the packet bus.

Packet Gateway (PGATE)—The PGATE board is used in newer systems; it connects the processor to the packet bus and terminates X.25 signaling.

Path Replacement—Path replacement is the process of rerouting an established call over a newer, more efficient path. The old call is then torn down, freeing up the resource. Path replacement is used on Q.SIG and DCS+ calls in conjunction with Look-Ahead Interflow (LAI) and Best Service Routing (BSR) to attempt to find the optimum path for an inbound call.

Path Replacement with Path Retention—When a call is transferred within a private network, the call's connection between the switches can be replaced with new connections while the call is active.

Permanent Virtual Circuit (PVC)—A PVC is a connection across a frame relay network, or cell-switching network, such as ATM. A PVC behaves like a dedicated line between source and destination end-points. When activated, a PVC will always establish a path between these two end points.

Personal Computer Memory Card International Association (PCMCIA)—The PCMCIA slot in a laptop was designed for PC memory expansion. NICs and modems can attach to a laptop through the PCMCIA slot.

Point-to-Point Protocol (PPP)—PPP is a protocol that allows a computer to use TCP/IP by the means of a point-to-point link. PPP is based on the HDLC standard that deals with LAN and WAN links, and operates at the Data Link Layer of the OSI model.

Port Network (PN)—PNs include both EPNs and PPNs. These two PNs differ in that the PPN contains an SPE. Both types of PNs consist of the following components: TDM bus, packet bus, port circuits, and interface circuits.

Priority Queue—The priority queue is a segment of a split's/skill's queue from which calls are taken first.

Processor Interface (PI)—A PI packet gateway (PGATE) circuit pack supports X.25 signaling between DEFINITY systems and adjuncts. The PI circuit pack is not included in new DEFINITY systems; however, PI channels are common in existing installations that still use X.25 connections.

Processor Port Network (PPN)—A PPN is required by each DEFINITY system; it contains the switch processing element (SPE), the system memory, the packet controller, and the network controller circuit packs.

Pulse Code Modulation (PCM)—PCM is a method of converting an analog voice signal to a digital signal that can be translated accurately back into a voice signal after transmission. A codec samples the voice signal 8,000 times per second, then converts each sample to a binary number that expresses the amplitude and frequency of the sample in a very compact form. These binary numbers are then transmitted to the destination. The receiving codec reverses the process, using the stream of binary numbers to recreate the original analog wave form of the voice.

Q.SIG—Also known as Private Signaling System Number 1 (PSS1), Q.SIG is an ISO standard that defines the ISDN signaling and control methods used to link PBXs in private ISDN networks. The standard extends the "Q" point in the ISDN logical reference model, which was established by the ITU-T in its Q.93x series of recommendations that defined the basic functions of ISDN switching systems. Q.SIG signaling allows certain ISDN features to work in a single- or multi-vendor network.

Queue—A queue is a collection point where calls are held until an agent or attendant can answer them. Calls are ordered as they arrive and are served in that order. Depending on the time delay in answering the call, announcements, music, or prepared messages may be employed until the call is answered.

Queue Directory Number (QDN)—QDN is an associated extension number of a split. It is not normally dialed to reach a split. The split can be accessed by dialing the QDN. The QDN is also referred to as a split group extension.

Reduced Instruction Set Computer (RISC)—The term RISC refers to microprocessors that contain fewer instructions than traditional CISC processors, such as Intel or Motorola. As a result, they are significantly faster. RISC processors have been used in most technical workstations for some time, and a growing number of PC-class products are based on RISC processors.

Remote Authentication Dial-in User Server (RADIUS)—RFC 2865 specifies RADIUS as a client/server network security protocol that performs remote use authentication and accounting. Remote users pass their credentials to the RADIUS server, which then either allows or denies network access.

Service Observing—Service observing is a feature used to train new agents and observe in-progress calls. The observer (split/skill supervisor) can toggle between a listen-only mode or a listen/talk mode during calls in progress.

Single Carrier Cabinet (SCC1) media gateway—An SCC1 consists of a single carrier containing the processor or IPSI circuit packs, tone-clock, port and service circuit packs, and a power supply. Up to four SCC1s may be stacked to form a PN.

Skill—Skill refers to the ability assigned to an agent to meet a specific customer requirement or call center business requirement.

Split—A split is a group of extensions/agents that can receive standard calls and/or special purpose calls from one or more trunk groups. Each split can be served by its own queue. If splits are defined according to agents' knowledge, they are called "skills." Depending on the ACD software, an agent can be a member of multiple splits/skills.

Spoofing—Spoofing means a router responds to a local host in lieu of sending information across a WAN link to a remote host. The local host thinks the response came from the remote host/network, when it really came from the router.

Split/Skill Administration—The split/skill administration refers to the ability to assign, monitor, or move agents to specific splits/skills. It involves changing reporting parameters within the system.

Split/Skill Supervisor—A split/skill supervisor is a person assigned to monitor/manage each split/skill and queue to accomplish specific objectives. A supervisor can assist agents on ACD calls, be involved in agent training, and control call intra-/interflow.

Subscriber Line Carrier (SLC)—SLC is a method of using T1 multiplexing technology to carry more lines over existing wires.

Subscriber Network Interface (SNI)—The SMDS protocol specifies how to connect CPE with an SMDS network. The point at which the CPE interfaces with the SMDS network is the Subscriber Network Interface (SNI). This interface, connecting a customer to the SMDS cloud, is usually implemented by means of T1 or T3 lines.

Switch Node (SN)—An SN reduces the amount of interconnect cabling between the PPN and EPNs by acting as a hub to distribute cabling. A system using a CSS can connect 3 to 43 PNs. A CSS can consist of up to three SN carriers.

Switch Processing Element (SPE)—When a telephone goes off-hook, or another type of device signals call initiation, the SPE receives a signal from the port circuit connected to the device. The SPE collects the digits of the called number, and the switch is set up to make a connection between the calling and called devices.

System Management Terminal—A system management terminal is the terminal from which system administration and maintenance is performed.

T1—T1 is one of the T-carrier telecommunication standards for multiplexing digitized voice signals. A T1 channel operates at 1.544 Mbps. Each T1 channel (64 Kbps) carries a digitized representation of an analog signal that has a bandwidth of 4,000 Hz. Originally, 64 Kbps was required to digitize a 4,000-Hz voice signal. Current digitization technology has reduced that requirement to 32 Kbps or less; however, a T-carrier channel is still 64 Kbps.

Tandem Switch—A telephone switch that forwards traffic between other switches is referred to as a tandem switch.

Threshold—A threshold is a point in time, or a criterion, that determines a certain action by a system. For example, the number of calls in a queue, or the time calls spend in a queue, can determine specific call treatments.

Time-Division Multiplexing (TDM)—TDM is a multiplexing technology that transmits multiple signals over the same transmission link, by guaranteeing each signal a fixed time slot to use the transmission medium.

Time-of-Day Routing—Time-of-day routing restrictions can be programmed into a PBX to limit outbound calls to certain hours of the day. This can be used to prevent unauthorized after hours use of the phone system by hackers.

TN2302AP IP Media Processor (Medpro)—The TN2302AP Medpro circuit pack provides packetized audio processing for an IP-enabled DEFINITY server or media gateway. The Medpro circuit pack supports the ITU-T H.323v2 recommendation and performs jitter buffering, echo cancellation, silence suppression, and DTMF tone detection. It supports the ITU-T G.711, G.723, and G.729 audio codecs.

TN2312AP IP Server Interface (IPSI)—The TN2312AP IPSI circuit pack provides call control messaging between Avaya media servers and IP-enabled port networks (PNs).

Transcoding—Transcoding is the process where a device converts a signal from one format to another. For example, an H.323 gateway transcodes packetized voice to analog for transport across the PSTN.

Translations—Translations are a set of programming instructions, stored within a CO switch, that defines the functions and services available on a line or trunk. The programming process is also called "translations," because it effectively tells a switch how to interpret, or convert, signals from other switches.

Trunk Group—A trunk group is a group of trunks that provide identical communications characteristics. Trunks within trunk groups can be used interchangeably between two communications systems or COs to provide multiaccess capability. An ACD has its own preassigned trunk groups.

Trunk State—The current status of a trunk is referred to as the trunk state.

Uniform (Universal) Dial Plan—Uniform dial plan is a private network numbering system that assigns unique extension numbers to each user, regardless of location. If locations in different cities are linked in a private telephone network, a uniform dial plan allows employees in different cities to dial each other using only extension numbers.

Vector Directory Number (VDN)—A VDN is an extension number used in ACD software to connect calls to a vector for processing. The VDN by itself may be dialed to access the vector from any extension connected to the switch.

Virus—A virus is a self-replicating, malicious program that spreads by attaching itself to a file. Viruses can spread quickly through a network, with effects that range from mildly irritating to highly destructive.

Voice over Internet Protocol (VoIP)—VoIP is a technology that consists of telephone signals transmitted as IP packets.

Wide Area Telecommunications Service (WATS)—WATS is a discounted long distance service. Customers can purchase incoming and outgoing WATS separately.

Worm—Worms, like viruses, are computer programs that replicate themselves and that often contain some functionality that interferes with a computer. The difference is that worms exist as separate entities; they do not attach themselves to other files or programs (like viruses). A worm spreads itself automatically over a network from one computer to another. Worms take advantage of automatic file sending and receiving features found on many computers.

Lesson 1—PBX Fundamentals

PBXs are telephone switches that are, in many respects, similar to central office (CO) switches. Like a key system, a PBX controls access to outside lines. However, a PBX makes the connections invisible to the user. In other words, a key system user must specifically select an outside line by pushing a button; a PBX user simply dials an outside call and the PBX system handles the details of connecting to an outside line.

Like CO switches, PBX systems are powerful, computerized systems that offer a wide range of sophisticated features. A PBX usually becomes cost-effective when a business grows to 20 to 50 lines. PBX systems are an economic necessity for businesses with thousands of lines in a campus environment.

Objectives

At the end of this lesson you will be able to:

- Describe the main components of a PBX

- Explain how a call is made through a PBX

- Define Direct Inward Dialing (DID)

- Name and briefly describe key PBX functions

- Name and briefly describe common auxiliary PBX equipment

 **Key Point**

PBX systems are similar to CO switches.

PBX Components

Significant differences exist among the features in PBX products, and in the way the features are packaged and sold. However, the general structure of most PBXs is similar, and includes the following major components, shown on the PBX Components Diagram:

- Trunks to connect the PBX to the telephone company

- Extension lines to connect the PBX to internal telephones

- Equipment cabinet: main processor, trunk cards, and line cards

- Telephones and attendant console
- Administrative terminal

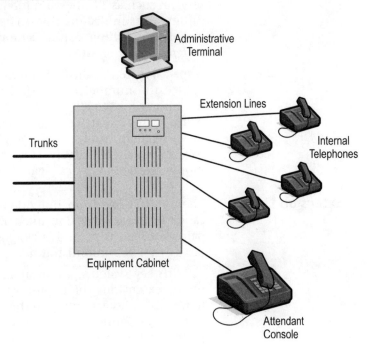

PBX Components

Trunks

The connections between a PBX and CO switch are called trunks. Trunks are usually large-capacity connections, such as T1 or Integrated Services Digital Network (ISDN) (we will explain both of these in Module 4). Each trunk can be configured for incoming calls, outgoing calls, or two-way calling.

Like any electronic system, a PBX requires a constant power source. If the system is not equipped with a backup power supply, certain trunks can be designated as Power Failure Transfer (PFT) trunks. If the power fails, these trunks are switched to standard, single-line telephones that draw their power from the telephone lines themselves (from the CO) and not from the PBX. This allows the business to continue answering incoming calls until power is restored.

Businesses configure their trunks depending on their telephone systems, business needs, and employee calling patterns. For example, a customer service center needs a higher number of incoming trunks, while a telemarketing business needs more outgoing trunks.

PBXs can connect to three types of external networks: local, inter-exchange, and private. In addition, many systems support special services such as T1 lines and foreign exchange (FX) trunks to local calling areas in distant cities. The variety of interface circuits is a key distinguishing feature among PBX products and different generations of products.

A special system, called a Subscriber Line Carrier (SLC) circuit, can reduce the number of physical wire pairs necessary to provide multiple trunks to a business. SLC uses multiplexing to allow one pair of wires to carry multiple call paths. SLC can provide from 2 to 96 separate trunks over a single physical connection. Because this approach effectively creates more connections, it is commonly referred to as "pair-gain" technology.

Extension Lines

The connections from individual telephone sets to the PBX are called lines, or extension lines. Each line is formed by a twisted pair of copper wire, and is assigned a telephone extension number by the PBX administrator.

As with key systems, a typical arrangement is to have more telephone extensions than trunks. However, the proportion of lines to trunks varies according to the needs of the business. Many customers configure their systems with 1 trunk for every 10 telephones, while businesses with heavy outside calling requirements may have 1 trunk for every 5 or fewer users. In general, however, most businesses need proportionally fewer outside trunks as the number of internal extensions grows.

Equipment Cabinet

The equipment cabinet, a large metal box in the main equipment room, is the heart of the PBX. It contains the computerized processors that perform telephone switching and other advanced features. The cabinet usually also contains the power supply, which converts 120-volt (V) AC current (standard "household" electricity) into the low-voltage direct current necessary to run the telephone system.

Much of the cabinet's circuitry is contained on electronic circuit cards that slide into slots and plug into a main processor board. Some of these cards allow the attachment of specialized accessory equipment. However, most of them are one of two main types:

- Trunk cards contain the circuitry necessary to communicate with the CO switch. Each trunk connects to a trunk card.

- Station cards contain the circuitry necessary to communicate with internal telephone extensions. Each line connects to a station card.

Trunks plug into one side of the equipment cabinet, attached to trunk cards. Lines plug into the other side, attached to station cards. The main processor board of the PBX makes connections between lines and trunks, and between lines. The portion of the processor that performs call switching is called the switching fabric.

Older PBXs are analog systems; most of today's PBXs are digital switches that use pulse code modulation (PCM) and time-division multiplexing (TDM) to switch voice calls within the switching fabric. Whether a PBX is analog or digital has significant implications for the connection between the customer's PBX and the CO, because an analog PBX supports only analog connections, while a digital PBX can support either digital or analog transmission service.

Telephones and Attendant Console

Telephone sets designed to work with a particular PBX system use digital signals and will not work when plugged into a "regular" analog telephone line (and in most cases, another vendor's digital line card ports). Each PBX vendor's telephone set includes programmable features that can be enabled or disabled by the company's telephone system administrator.

On the other hand, though digital PBXs support analog telephone sets connected to analog line cards, analog sets will not work when plugged into a digital line card. Many analog modems have been irreparably damaged by the voltages present on the digital line card ports.

The attendant console, used by the office receptionist or a dedicated PBX operator, is essentially a larger telephone set. This console contains features and controls that perform the specialized operator-assisted functions we will discuss later.

Administrative Terminal

A computer workstation provides administrative access to the PBX system. The company's telephone administrator uses the system to configure the wide range of features and options available.

Making a Call on a PBX

Making a telephone call within a PBX environment is very similar to making a call from home using the public telephone network.

When you lift the receiver, placing the telephone in the off-hook position, the telephone signals the PBX that it needs a connection. The PBX then sends dial tone to the extension telephone, and waits for incoming touch tones.

To call another extension within the same PBX system, the caller enters the digits of the extension number. When the PBX detects the tone that represents the first digit, it removes dial tone from the line. When the complete extension number has been entered, the PBX sends a ringing signal to the called extension, and connects the two extensions when the called party answers.

To make an "outside" call, the caller must first signal the PBX, usually by dialing 9. This special number tells the PBX to seize an available outgoing trunk for the call. The PBX then signals the telephone company's CO switch to send dial tone. The second dial tone confirms that the caller is now connected, through the PBX, to the public telephone network. The caller then enters the destination telephone number, and the call follows the process described in Module 1.

Call Routing

The primary function of a PBX is to provide internal access to telecommunications trunks provided by a telephone company. A PBX can also provide access to other PBXs (such as other company locations) by means of dedicated tie lines.

If necessary, a PBX can queue or hold calls until a trunk becomes available. PBXs can also provide custom routing options for each call, based on the called party number, caller's location, and other parameters. In other words, the PBX reviews these factors and decides whether to route the call over the internal network, a discount long distance provider, or a specialized data service.

A feature called Automatic Route Selection (ARS) allows the telecommunications manager to program different trunks for use based on time of day, type of call, or any number of other factors. ARS is becoming increasingly important as more customers send voice calls by means of the Internet or public data networks.

Supplementary PBX Features

In addition to its core system features, a PBX can provide many supplementary services:

- **Call hold**—Callers can be placed on hold, or in parking orbits where the call is left until the called party retrieves it by means of an access code.

- **Music on hold**—While callers are waiting, they can hear music or messages about new products and services. TechData, a national computer distributor, even sells on-hold time to companies in the form of advertising.

- **Hunting**—Incoming calls can be routed to other extensions or lines if the first destination is busy. Hunting is commonly used in customer service centers and help desks to distribute incoming calls among a group of representatives (called a "hunt group").

- **Call restrictions or "class of restriction"**—A PBX can be programmed to restrict calls in various ways. This is usually done to prevent an extension, such as a lobby courtesy telephone, from making long distance calls. You can program a PBX extension down to the smallest detail, even restricting access to specific numbers.

- **Call tracking**—By requiring callers to enter a numerical code, many aspects of a call can be tracked. For example, law firms use this feature to track the duration of calls attorneys make on behalf of their clients (for later billing).

DID

Direct Inward Dialing (DID) connects incoming callers directly to the employees they need to talk with. DID is an effective means of increasing customer contact while reducing the load on PBX operators.

In a traditional PBX, the console operator(s) or automated attendant must answer all incoming calls and connect them to destination extensions. The Incoming Call Diagram shows an outside caller placing a call over a standard inbound trunk. The attendant answers the inbound call, and connects the call to the destination user. However, with a PBX equipped for DID, coupled with DID service from the telephone company, incoming calls can be directly routed to individual extensions or departments served by the PBX.

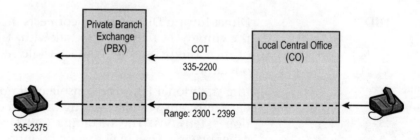

Incoming Call

The DID Trunk Diagram illustrates a carrier's switch providing DID service to a customer's CO. The CO switch recognizes the called number as falling in that customer's defined DID range, routes the call to the appropriate DID trunk, seizes the trunk, and passes only the last four digits to the PBX. The PBX then routes the call to the appropriate subscriber extension.

DID Trunk

DISA

A company can significantly reduce its telephone expenses by selecting the right long distance carrier. However, it is often difficult for outside employees, such as salespeople or field service representatives, to access the low-priced carrier for work-related calls.

Direct Inward System Access (DISA) allows an outside caller to dial directly into the PBX system, then access the system's features and facilities remotely. DISA is typically used to allow employees to make long distance calls from home or any remote area, using the company's less expensive long distance service. To use DISA, an employee calls a special access number (usually toll-free), then enters a short password code.

Although DISA is a great convenience to remote employees, unauthorized people often acquire the toll-free access number and use it to steal long distance telephone service. Changing the password codes from time to time can help prevent this. Still, it is best if the telephone administrator restricts DISA to only those employees who require it, and regularly checks the PBX call records to detect any unusual patterns.

Station Set Features

A PBX telephone set includes a large selection of features. Like programming a CO switch, programming the lines and features on each telephone set can be a complicated task.

There are literally hundreds of features in most PBX systems; however, most users cannot remember how to use more than two or three of the most common ones. Traditionally, a user must briefly press the telephone switchhook to access the set's features. However, to make these features easier for users to understand and remember, many telephone sets now include special function keys to access popular functions such as:

- **Call forwarding**—An extension can be programmed to send incoming calls to another extension.

- **Call transfer**—After answering a call, an employee can transfer it to another extension.

- **Speaker paging**—By using an access code, employees can broadcast announcements from a telephone extension. Paging is commonly used in medical centers, manufacturing locations, and retail stores.

- **Speed dialing**—Commonly used telephone numbers can be programmed as abbreviated two- or three-digit access codes.

- **Call pickup**—If one person is away, any other person from the same defined workgroup (called a "pickup group") can enter an access code and answer a call that is ringing at the absent person's desk.

- **Message waiting light**—A light on the desk set glows when the user has voice mail. If a desk set does not have a message light, a "stutter tone" can alert the user to new voice mail. The user hears the tone when the message comes in during a call, or when the user picks up the phone to make a call.

Some PBX telephone sets, such as those for hotel/motels, include special features for their increasingly telephone/computer-intensive guests. These sets offer two lines, lots of special features, and voice mail.

Attendant Console Services

PBX operators are often the most vital part of a company's information system: they usually know where everyone is, who has moved where, and what to do in case of an emergency. In addition, they can provide the following key functions from the PBX operator (attendant) console:

- **Call processing**—A PBX operator can connect incoming callers to PBX extensions.

- **Company directory assistance**—A PBX operator can provide extension numbers and direct-dialing numbers.

- **Organizing "meet-me" conference calls**—In a meet-me conference, callers dial a predetermined telephone number and security access code, which connects them to a telephone conference call that may include up to six people. In some cases, the conference connections are made automatically; in others, the PBX operator serves as the "conference bridge," greeting incoming callers and connecting them to the conference.

- **Night service**—When PBX operators go home and turn off the console, incoming calls can be directed to ring throughout the building. From any extension telephone, an employee can dial an access code to answer the incoming call.

- **Camp on**—A PBX operator can place an incoming call on hold, waiting for a specific extension that has a call in progress. As soon as the first call is completed, the camped-on call rings at the extension.

Administration and Management Reports

As we have seen thus far, the PBX administrator plays a very important role, configuring the system and ensuring it is used correctly. The following management features provide detailed information on calls being processed and completed through a PBX system.

Call Detail Reports

Most telecommunications managers do not consider themselves telephone police; however, companies do consider unauthorized telephone abuse to be theft of company property. Telephone fraud represents a $1 billion loss to American businesses each year.

To help a PBX administrator prevent this sort of fraud, most PBX systems have the ability to record all calls and sort them by almost any criteria, including date, time, number called, frequently called numbers, extension number, and user name. If necessary, a PBX report can identify a specific extension that called a particular number at a certain time.

Hotel/motel PBXs have the ability to print a Call Detail Report instantly, to total the telephone-use fees that must be added to a customer's bill at check-out time. Without this instant reporting, hotels could lose thousands of dollars of revenue.

Test Reports

PBXs, like CO switches, perform a variety of internal tests. The results of these tests are presented as reports, often in Hypertext Markup Language (HTML) format for viewing with an Internet browser. These reports help the administrator monitor the general health of the system, and are useful to service technicians in case of a failure.

Busy-Hour Studies

One of the administrator's key responsibilities is to determine the number and type of trunks a business needs. Special management reports, called busy-hour studies, provide information that helps an administrator make those decisions.

The "busy hour" is the hour during which a network or office telephone system carries its greatest traffic. For most businesses, the busy hour occurs in the mid-morning, as employees organize their work for the day. The general objective of an administrator is to provide enough transmission capacity to carry busy-hour traffic.

In actual practice, it is wasteful and expensive to provide enough capacity to carry 100 percent of the busy-hour traffic. Thus, an administrator must determine how much of this peak traffic is acceptable to block. In other words, the administrator estimates the acceptable percentage of inbound callers who will receive a busy signal, and the percentage of outbound callers who must wait for a free line.

PBX Enhancements

A variety of other systems can be added to a PBX to provide specialized and enhanced services.

Voice Response

Many PBXs shipped today come with voice mail and voice response, also known as Integrated Voice Response (IVR) or Voice Response Unit (VRU). These systems allow callers to use the telephone touch pad to access information ranging from store locations to credit card transactions. Some banks even allow customers to pay their bills by means of IVR.

IVR is a rapidly growing field, which can reduce the staff workload and, when properly implemented, increase the level of customer service. When a call center needs to support self-service applications and calls seeking information only, an IVR allows the center to handle increased call volume without increasing agent staffing levels.

Voice Recognition

Voice recognition technology allows a computer to understand the human voice. This technology has been around for more than two decades, and it is widely used as a cost saver. Computer answering services now ask you to speak your instructions, and directory assistance services ask you to speak the name and location of the number you need.

As 10-digit dialing becomes more common in large cities, voice recognition may be more accurate than button pushing for entering telephone numbers. However, one of the greatest potential applications is voice Internet access, providing access to electronic mail (e-mail) and other applications for people without personal computers (PCs).

Call Following

Users want calls to get through to them, wherever they are, whatever the time of day. As staff increasingly work remotely, calls can be forwarded to a sequence of different locations until the called party is found, as shown on the Follow-Me Call Flow Diagram.

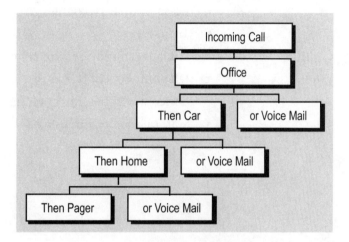

Follow-Me Call Flow

ACD

Customer service centers that receive a high volume of incoming calls use a special kind of PBX called an automatic call distributor (ACD). ACDs route incoming calls to agents based on programmed criteria, such as the agent that has been idle the longest. If an agent is not available to take a call, the ACD holds the call in a queue and the caller hears a "please hold" message.

Other Auxiliary Equipment

Other common types of auxiliary equipment include:

- **Station Message Detail Recorder (SMDR)**—Creates a log of all incoming and outgoing telephone calls

- **Uninterruptible power supply (UPS)**—Enables a PBX to operate during a power failure

- **Direct Station Selector (DSS)**—Displays which stations are being used in a telephone system

- **Headset**—Provides hands-free telephone operation for either analog or digital telephones

PBX Costs and Requirements

In addition to the one-time installation cost, a PBX system usually incurs the following ongoing charges:

- Equipment maintenance, upgrades
- Lease or finance charges
- Trunk usage
- DID termination service charge
- DID numbers in blocks of 10, 20, or 100
- Office space, heating/cooling, electrical power
- Staff salaries and training

Activities

1. Explain how a PBX processes a call.

    _____

    _____

    _____

    _____

2. Match the PBX functions and features to its description.

    ARS

    Call tracking

    Call forwarding

    DID

    Hunting

    Call pickup

    Music on hold

    DISA

    Call restrictions

    a. Callers hear music or messages while waiting for the destination extension to pick up _____

    b. Controls what types of calls may be placed from an extension _____

    c. Allows users to dial in to the PBX to access inexpensive long distance capabilities _____

    d. Allows callers to bypass the attendant or operator and ring the destination extension directly _____

    e. A capability programmed into an extension that transfers calls made to one extension automatically to another _____

    f. A feature that allows any one of a defined workgroup of extensions to answer a call destined for any other workgroup member _____

    g. Provides the capability to record call aspects through the use of numeric codes _____

    h. Routes calls to alternate extensions if the original destination is busy _____

    i. Allows control of trunk use based on factors such as time of day or call type _____

3. A PBX auxiliary function that provides callers the ability to access company information by means of their keypad is known as _____.

    a. IVR

    b. Voice recognition

    c. Call following

    d. ACD

4. _____ is a PBX function that routes incoming calls to agents based on programmed criteria.

    a. IVR

    b. Voice recognition

    c. Call following

    d. ACD

5. A PBX function that allows company staff to effectively work remotely by forwarding calls to them wherever they are located: _____.

    a. DSS

    b. Call following

    c. ACD

    d. Camp on

Extended Activities

1. Many vendors provide PBXs for small to very large companies. Some of the larger players are Avaya Communication (formerly Lucent Technologies) and Nortel. Research the products available from these companies, and compare their features and applications.

2. Another feature of PBXs, commonly used in telemarketing firms, is the autodialer. The autodialer automatically dials telephone numbers in a database, and when the called customer picks up, connects the call to the first available telemarketing representative.

    Listen carefully when the next telemarketer calls your home or business. You may hear a pause before someone answers your "hello." This is the delay between the time you answer the call and the time it takes the autodialer to connect to a representative. A properly programmed autodialer will provide no discernible pause.

Lesson 2—PBX Features and Functions

A PBX can improve the operation of many businesses. However, some businesses cannot do without a PBX, because they rely on specific features of PBX systems. This lesson reviews many of the features and functions found with typical PBX systems and discusses how they are typically used.

Objectives

At the end of this lesson you will be able to:

- Describe what a call accounting system is and what it does

- List and describe the primary PBX features

- Give an example of how each PBX feature can be used in a typical business

 **Key Point**

A PBX can do almost anything a CO switch can do.

Call Accounting

Call accounting systems incorporate SMDR (also known as Call Detail Reporting [CDR]) systems. Call accounting systems are computers that track individual telephone calls. Call accounting systems are often dedicated PCs connected to a PBX by means of a serial port. The PC and associated software gather information about every telephone call processed by the PBX, and stores information about each call, such as call duration, source, destination, and time placed. Call accounting software is used to retrieve the information, sort it, and provide reports about an organization's calling patterns. The Call Accounting Configuration Diagram illustrates this type of system.

Call accounting systems are used for more than simply tracking calls and producing managerial reports. They are often used to track the usage of lines and trunks for traffic engineering purposes.

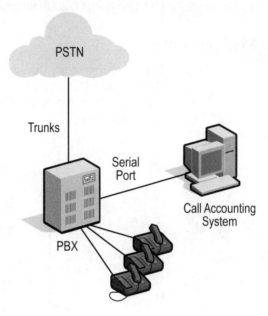

Call Accounting Configuration

Authorization Codes

Call accounting systems may also detect fraudulent use of telephone systems. If a company thinks the telephone system is being used inappropriately, authorization codes are issued to users of the system. These codes are used to create reports that detail the telephone calls for individual departments or employees. In addition, authorization codes can allocate, or "bill-back," the cost of telephone service to each department's budget.

Account Codes

Call accounting reports are essential to some types of businesses. For example, law and accounting firms use detail reports to bill clients for telephone time spent on their behalf. These types of organizations use account codes coupled with call accounting systems. Each customer has a specific account code, so the call accounting system can produce detailed reports of the telephone usage for each account.

An alternative to on-site call accounting hardware and software is a service bureau. Large companies often hire service bureaus to administer call accounting services from a remote location. Service bureaus dial into the PBX from a remote location and collect the client organization's call data.

Voice Mail Systems

Voice mail systems are answering machines built into a PBX, or stand-alone PCs with voice mail software. A configuration showing a PC-based voice mail system is shown on the PC-Based Voice Mail Diagram.

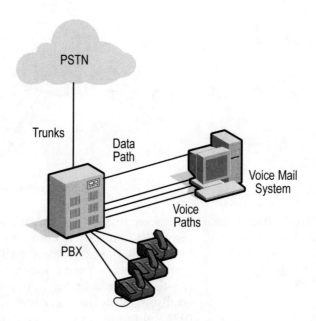

PC-Based Voice Mail

With PC-based voice mail systems, two types of connections are made between the PBX and PC. Voice paths connect an outside caller with the mailbox of the called party.

A data path is used to carry call control information, such as the dialed number or the caller ID (for internal calls or ISDN), and control signaling that enables features such as call notification lights.

Signaling information may be passed between the voice mail system in the PBX over analog or digital voice lines. Other ways of passing signaling information include RS-232-C and packet switched serial connections between devices.

When an inbound caller receives a busy or no answer condition, the PBX sends the call to a hunt group programmed for the voice mail system. The PBX and its software locate a free line within the hunt group and connect the call to the voice mail system. If for some reason the voice mail system fails, the system administrator may reprogram the PBX to route calls to an attendant or to disable the voice mail hunt group temporarily.

If the system uses in-band signaling, call information travels across either the same line or over a second analog line. A system that uses a digital link such as Station Message Desk Interface (SMDI) passes call information over this link. The signaling information tells the voice mail system which mailbox to open. The analog connection then passes the mailbox owner's greeting to the caller, and carries the caller's message to the voice mail system where it is recorded on the storage medium.

When the caller hangs up, the PBX signals the voice mail system to release the line. The voice mail system then sends a signal to the PBX to turn on the message-waiting indicator.

FRL

When you set up a Class of Restriction (CoR), you specify a facility restriction level (FRL) on the CoR screen. The FRL sets up the calling privileges of the user. FRLs range from 0-7. For the CoR assigned to a station, FRL 7 allows the highest level of calling privileges, while FRL 0 represents the lowest level of calling privileges. You also assign an FRL to each route pattern preference in the route pattern screen. For the FRL assigned to a route pattern, FRL 7 has the highest level of restriction, while FRL 0 is the lowest level restriction. For example, you want a station to be able to call international, toll free, and local; and the station CoR has an FRL of 6. The route pattern would have to have an FRL of 6 or lower to complete the call; an FRL of 7 would not allow the call to go through. When a call is made, the system checks the user's CoR. The call is allowed if the caller's FRL is higher than or equal to the route pattern's FRL.

Design Factors

When implementing a new voice mail system, you must consider the PBS software features, the number of system users, and the voice messaging ports on the PBX. In other words, the PBX must support voice mail and have enough voice messaging ports to accommodate the maximum number of simultaneous voice messages that could potentially be placed.

You must also decide how to notify users that voice mail is waiting. In most cases, this will be determined by the features available on existing desk sets. If existing phones have "message waiting" lights, most users prefer to use them. If lights aren't available, and the customer does not want to upgrade the equipment, then the system can deliver a "stutter tone" to alert users to waiting messages. The user hears the tone when the message comes in during a call, or when the user picks up the phone to make a call.

Finally, to store user messages, you must determine how much storage space the system will need. Different types of users work with messages in different ways. For example, an office worker would likely access his or her voice messages several times a day, either deleting or storing them as needed. This regular pattern ensures that messages that have been heard would be removed from system storage on a daily basis.

However, traveling users may access their voice messages only a few times a week. This means that their messages will likely use more storage space than the office worker's. It is therefore important to consider the number and types of users a voice mail system supports to determine the appropriate amount of message storage space needed.

Unified Messaging

Some voice mail systems also come with unified messaging features. Unified messaging systems provide a single graphical interface that helps users manage voice mail, e-mail, and faxes.

Incoming messages are stored on the voice mail system's hard drive, or if integrated with an e-mail system such as Microsoft's Exchange Server, based to the e-mail server for delivery to the user. Notifications are sent, by means of a local area network (LAN), to each user's computer. Using the graphical interface, users can read text documents, listen to voice mail (on computer speakers or the telephone), store documents on the local hard drive, or print text messages.

Digitized voice messages use a considerable amount of network resources, well beyond that used by a traditional text-based e-mail. If the messaging server is not equipped with enough resources (disk space, CPU, memory, network bandwidth, and so forth) to adequately support voice messaging, server performance will deteriorate, users will experience delays in accessing their messages, and user mailboxes will quickly fill to exceed their allocated storage space.

Broadcasting

Voice mail systems also offer message broadcasting, which sends the same message to multiple voice mail boxes.

Voice Mail Security

In Lesson 1, you learned that Direct Inward System Access (DISA) can be a target for toll fraud when improperly administered. Advanced voice mail features also have the potential to become a security problem if not carefully set up and closely monitored.

In general, voice mail is vulnerable to several types of potential security problems:

- **Curiosity**—Hackers motivated by the excitement of breaching a system. They may not intend to do harm, but may damage the system as they explore it.

- **Malice**—Intentional damage done by disgruntled employees, ex-employees, or anyone else who dislikes the company. Depending on the size and number of voice mail security holes, this damage can range from rude messages left in mailboxes to severe system disruptions.

- **Industrial espionage**—Voice mail is often full of confidential messages, sales leads, and other sensitive information that could damage the company or an employee if known by a competitor.

- **Illegal business use**—Some criminals breach voice mail systems to set up untraceable mailboxes for their own use.

- **Toll fraud**—Criminals use advanced features, such as call transfer or outcalling, to access outgoing lines and steal long-distance service. This is a popular tool among drug dealers who regularly call South American or Asian countries.

As with DISA, prevention is the best way to protect a voice mail system. The following security guidelines will discourage most hackers. Many of these are already implemented in major voice mail systems:

- Grant system privileges only as needed. Use Class of Service restrictions to customize each user's access to features such as remote access or call transfer.

- Prevent callers from transferring from voice mail to an outgoing line, or restrict outbound transfers to certain numbers.

- Eliminate all unused voice mail boxes. Do not set up boxes before they are needed, and close a box as soon as an employee leaves the company.

- Be serious about passwords. When assigning a new mailbox, create a random password. To make passwords harder to guess, require them to be at least six to eight characters long; set even longer passwords for administrative access. Require users to change their passwords regularly, and emphasize the importance of not writing passwords in obvious places.

- Limit the number of attempts to log in to voice mail. Some systems lock a voice mail box after three unsuccessful password attempts.

- Regularly review reports of system activity for unusual patterns.

- Plan for the worst. Voice mail security should be a key element in your overall communications security plan. This plan should include predetermined procedures to follow in case of a security attack. As you develop your plan, make your telecom service provider a full partner in the process. By including your provider in your procedures, you can more quickly shut down a security attack if one occurs.

Automated Attendants

Automated attendants are machines, integrated into PBX systems, that answer the telephone and play a now-familiar recorded message:

> "Thank you for calling our company. If you know your party's extension, please dial it now. Otherwise, stay on the line and an operator will assist you shortly."

Automated attendants can supplement a staff of live operators during very busy times, when the staff may not be able to handle all of the incoming calls. Some businesses use an attendant to answer all incoming calls, allowing a receptionist to perform other tasks when not providing personal service to callers. Automated attendant systems can also allow callers to select options from a voice menu, to automatically route calls to different departments.

CoR

Class of Restriction (CoR), also called Class of Service (CoS), allows a telephone administrator to define user rights for a telephone system. Some PBXs allow an administrator to define up to 64 classes of telephone users. Each of those classes can be assigned a different set of call-origination and call-termination restrictions. After classes have been defined, the administrator assigns individual users to groups, as well as voice terminals, voice-terminal groups, data modules, and trunk groups.

A simple example of CoS is a telephone that does not permit direct long distance calls, such as a lobby courtesy telephone or hospital room. In large organizations, CoS can be used to specify whether voice-terminal users can activate any number of PBX features, or access certain trunks or services. For example, a class of service can allow users to make long distance calls, but only after they enter an account code to authorize the call and track the toll charges. Other CoS settings can make it impossible for users in certain departments to call each other.

Call Pickup

Call pickup allows someone to remotely answer the ringing telephone of a person who is not available. To use call pickup, simply dial a code number (or press a function key) plus the extension that is ringing, then answer the call as if it were your own.

Call Park

Call park is a more flexible type of call hold, that allows the call recipient to resume the call from any extension on the system.

Call Forwarding

Call forwarding allows a telephone user to redirect a telephone call to another telephone. For example, a telephone in an office can be forwarded to an extension in another part of the building.

Call Announcement

When a PBX user transfers a call to another user, call announcement allows the first user to speak to the second user before the second person answers the forwarded call. This feature makes smoother customer service possible, because the first user can explain who the caller is, and what has transpired, before the second user takes over the call.

In contrast, a "blind transfer" means a call is simply forwarded to a second extension. The first user cannot control the call or talk to the new called party.

Call Waiting

Call waiting alerts a called party that another incoming call is waiting. Without call waiting, the second caller receives a busy signal. With call waiting enabled, the called party hears a short beep or click when a second call comes in. To put the first caller on hold and answer the second call, the called party briefly presses the switch hook once.

DAU

When customers are waiting in a queue to talk to an agent, they will wait more patiently if they are politely encouraged to do so. A digital announcement unit (DAU) plays a variety of recorded messages to callers on hold, according to the call treatment script. For example, a DAU typically plays an initial greeting that asks customer to stay on the line and wait for the first available agent. Subsequent messages can reassure callers that they will soon be served, or invite them to exit the queue by leaving voice mail messages for agents.

Facility Test Call

Facility test call is a system feature that allows administrators to test four types of system facilities. The four testable facilities are:

- **Trunk test call**—Accesses and tests tie or CO trunks, but not test DID trunks

- **Touch-tone receiver test call**—Accesses and tests the DTMF receivers on a Tone Detector or Call Classifier circuit pack

- **Time slot test call**—Connects to a specific time slot in the TDM bus·

- **System tone test call**—Connects to a system tone

Activities

1. Match the following terms with the correct description:

   Also called CoS

   Telephone call redirection

   Machines that answer telephone calls

   Used for telephone call accounting

   Remote answer of a call

   Recorded message playing while caller is on hold

   Resume a call from any extension

   Computers that track telephone calls

   Incoming call alert

   a. Call accounting

   _____

   b. Authorization code

   _____

   c. Automated attendant

   _____

   d. CoR

   _____

   e. Call pickup

   _____

   f. Call park

   _____

   g. Call forwarding

   _____

   h. Call waiting

   _____

   i. DAU

   _____

2. What is a facility test call used for?

   a. To connect to a system tone

   b. To connect to a specific time slot in the TDM bus

   c. To access and test tie or CO trunks

   d. All of the above

Extended Activity

1. Go to the following Web site (**http://www1.avaya.com/ enterprise/az.html**) and find information on the following subjects:

   a. Call accounting systems

   b. Voice mail systems

   c. Automated attendants

   d. CoR

   e. DAU

Lesson 3—PBX Configurations

This lesson provides an overview of Avaya Communications Manager (ACM) systems (formerly DEFINITY ECS). It is a general overview of product capabilities. The features described here are specific to product models and release levels.

Objectives

At the end of this lesson you will be able to:

- Name and describe the main components of an Avaya DEFINITY Server with ACM system

- Draw and explain the main DEFINITY Server with ACM configurations

- Describe the basic ACM software architecture

 **Key Point**

Avaya ACM systems process both traditional and IP voice and data traffic.

Overall Connectivity

The PBX Connectivity Diagram illustrates how a PBX connects externally to the public-switched telephone network (PSTN) and internally to different types of terminals. The PBX handles incoming and outgoing calls (traffic flow) by switching inbound and outbound calls to the appropriate ports.

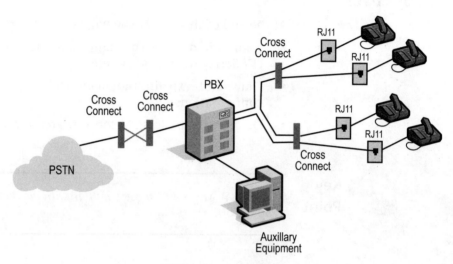

PBX Connectivity

Cross connects are used to connect one circuit (physical path) to another. They are used to provide for easy configuration and rearrangement of physical wiring. Wires are "punched down" onto a termination block that has many connections available for each pair of incoming and outgoing wires. Extra connections are available on the termination block for jumper wires, used to connect pairs.

Avaya DEFINITY Server with ACM System Components

Avaya DEFINITY Server with ACM is a connection-oriented digital switch that processes and routes voice communications (telephone calls) and data communications from one endpoint to another. The DEFINITY Server with ACM Diagram illustrates this type of system.

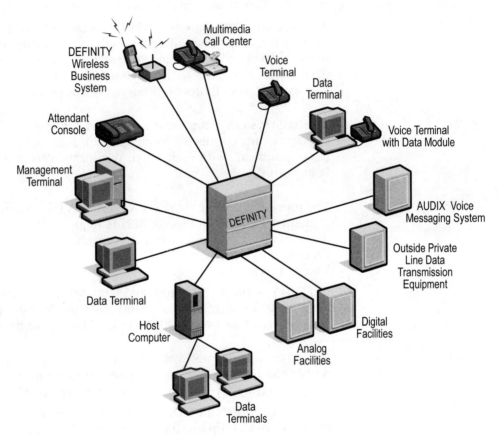

DEFINITY Server with ACM

Digital Networking

Within the DEFINITY Server system, both voice and data are processed as digital signals. When analog calls enter the system from outside, the DEFINITY Server converts them to digital signals; signals that are already digital do not need to be converted. Outgoing digital signals from the system are converted to analog signals for the analog lines and trunks. This type of dual-purpose system is often called a converged network, or digital networking.

As voice and data signals travel between endpoints, the signals enter and leave the switch through port circuits. The switch makes high-speed connections between analog and digital trunks, data lines connected to host computers, data-entry terminals, PCs, and Internet Protocol (IP) network addresses.

PN

The basic system component is the port network (PN), consisting of port circuits connected to internal buses that allow the circuits to communicate with each other. We discuss PNs in greater detail later in this lesson.

PPN

The processor port network (PPN) contains the switch processing element (SPE); therefore, one PPN is required. The SPE is a computer that operates the system, processes calls, and controls the PN containing the port circuits. We discuss the SPE in more detail later in this lesson.

DEFINITY Servers converted to S8700 Media Servers no longer have PPNs. Existing PPNs are converted to PNs by replacing the control circuit packs with IP Server Interface (IPSI) circuit packs.

EPN

An expansion port network (EPN) (optional) contains additional ports that increase the number of connections to trunks and lines. Unlike a PPN, an EPN does not contain an SPE.

Each EPN can extend PBX connectivity to remote locations. For example, if a company has one home office and three branch offices, each branch would install an EPN as a remote shelf. In each branch, all telephone devices connect to the EPN; each EPN connects to the home office PPN.

CSS

A center stage switch (CSS) is the central interface between the PPN and EPNs. The CSS serves as a hub for wide area network (WAN) carriers that connect multiple locations to the same central switch.

A CSS consists of one, two, or three switch nodes (SNs). A CSS connects to each EPN using fiber optic facilities. These fiber links can be public WAN services such as Asynchronous Transfer Mode (ATM) or T1, or dedicated private links that connect campus buildings.

SN

A switch node (SN) reduces the amount of interconnect cabling between the PPN and EPNs by acting as a hub to distribute cabling. A system using a CSS can connect 3 to 43 PNs. A CSS can consist of up to three SN carriers.

- One SN supports 1 to 15 EPNs.
- Two SNs support up to 29 EPNs.
- Three SNs support up to 43 EPNs.

Note: The number of EPNs that can be connected with two or three SNs may be less, depending on the internal SN traffic.

Each SN contains 1 to 16 switch node interface (SNI) circuit packs. Each interface can connect to a PN or another SN using fiber optic cable. One interface always connects to the PPN and one connects to each EPN. The Switch Node Diagram shows the CSS linking the PPN to EPNs by the SNI circuit packs in an SN carrier.

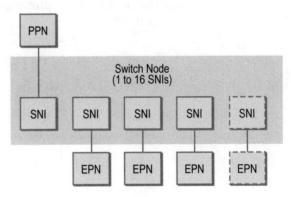

Switch Node

In a high-reliability system (with a duplicated processor), the CSS can also consist of two, four, or six SN carriers (duplicated SNs). Two SNI circuit packs connect to the PPN, allowing up to 15 PNs to connect to 1 SN, up to 29 PNs to connect to 2 SNs, and up to 43 PNs to connect to 3 SNs, depending on the exact configuration chosen.

ATM SN   Any standards-compliant ATM switch can serve as a DEFINITY ECS SN. In this configuration, TN2305 multimode or TN2306 single-mode ATM circuit packs are installed on the port networks and connected to the ATM switch with the multi-or single-mode fiber specified for the ATM switch.

Main System Configurations

The Main System Configurations Diagram shows the following five main system configurations:

1. **Basic system**—A PPN only.

2. **Directly connected system with three PNs**—One PPN is directly connected to one or two EPNs, and the EPNs are connected directly to each other.

3. **CSS-connected system**—One SN connects up to 15 EPNs to the PPN.

4. **CSS-connected system with two or more SNs**—Two SNs connect up to 29 EPNs to the PPN. Three SNs connect up to 43 EPNs to the PPN.

5. Up to 43 EPNs connected by an ATM switch.

6. Up to 43 EPNs connected over a WAN with multiple ATM switches.

Note: A system cannot connect over ATM and to a CSS simultaneously.

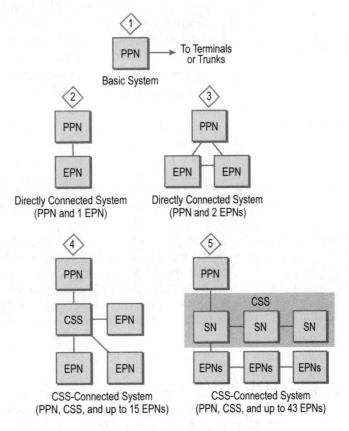

Main System Configurations

PN Components

PNs include both EPNs and PPNs. These two PNs differ in that the PPN contains an SPE. Aside from the SPE (discussed later in this lesson), both types of PNs consist of the following components:

- TDM bus
- Packet bus
- Port circuits
- Interface circuits

TDM Bus

The Time-Division Multiplexing (TDM) bus runs internally throughout each PN, and terminates on each end. It consists of two 8-bit parallel buses, Bus A and Bus B, that are normally active simultaneously. These 8-bit buses carry switched digitized voice, data, and control signals among all port circuits, and between port circuits and the SPE.

Each TDM bus has 484 time slots. It conforms to the ISDN architecture, with 23 B channels and 1 D channel available per bus.

Packet Bus

The packet bus also runs internally throughout each PN, and terminates on each end. It is an 18-bit parallel bus that carries logical links and control messages from the SPE, through port circuits, to endpoints such as terminals and adjuncts. The packet bus carries X.25 links, SNIs, and remote management terminal traffic. The DEFINITY G3csi systems do not support the packet bus.

Port Circuits

Port circuits form analog/digital interfaces between PNs and external trunks and devices, and provide links between these devices and the TDM bus or packet bus. Port circuits convert incoming analog signals to PCM digital signals, and place them on the TDM bus. Port circuits also convert outgoing signals from PCM to analog for external analog devices.

All port circuits connect to the TDM bus. Only specific ports connect to the packet bus.

Interface Circuits

Interface circuits, located in the PPN and in each EPN, are port circuits that terminate fiber optic cables. Fiber optic cables are used to connect the TDM and packet buses in the PPN cabinet to the TDM and packet buses in each EPN cabinet. Fiber optic cable also connects the CSS to the PPN and EPNs.

These interface and cabling terminations provide a transmission path between the port circuits in different PNs.

- Digital Signal Level 1 (DS1) (T1) interface circuits convert fiber interface to DS1 interface between PNs for DS1 remoting.

- Service circuits connect to an external terminal used to monitor, maintain, and troubleshoot the system. They also provide tone production and detection, as well as call classification, modem pooling, recorded announcements, and speech synthesis.

- ATM interfaces interconnect ATM PNs

The Directly Connected System Diagram shows how the major port network components interconnect in a simple configuration. Here, a PPN is directly connected to two EPNs by fiber optic cable. The EPNs are also connected to each other by fiber optic cable. This approach corresponds to configuration #3 in the Main System Configurations list above.

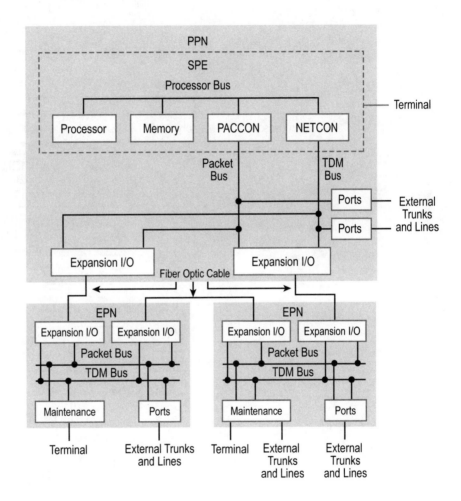

Within the PPN and EPNs, TDM and packet buses route voice and data calls between external trunks and lines.

The CSS-Connected System Diagram shows a configuration that uses a CSS to route voice and data calls between external trunks and lines.

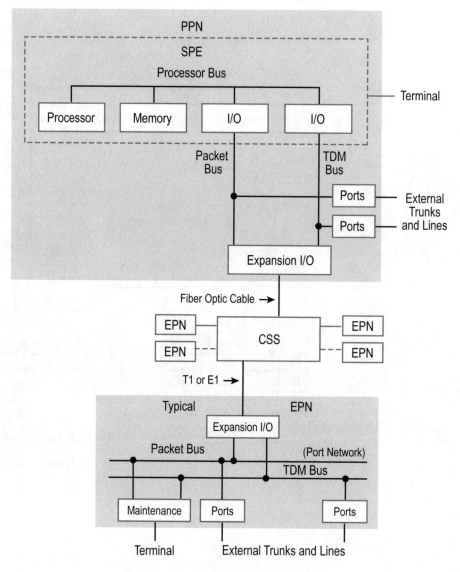

CSS-Connected System

SPE

When a telephone goes off-hook, or another type of device signals call initiation, the switch processing element (SPE) receives a signal from the port circuit connected to the device. The SPE collects the digits of the called number, and the switch is set up to make a connection between the calling and called devices.

The SPE consists of the following control circuits connected by a processor bus:

- **Processor**—All systems use a reduced instruction set computer (RISC) processor. Several processor circuit packs are used:
    - TN2402 on the DEFINITY Server CSI.
    - TN2404 on the DEFINITY Server SI.
    - TN2314 on the S8100 with Compact Modular Cabinet (CMC1) (formerly DEFINITY ONE), S8100 with the G600 media gateway (formerly IP600), and the S8300/S8500/S8700 Media Servers.
    - UN331C on the DEFINITY Server R.
- **Memory**—The different systems use a variety of memory configurations:
    - DEFINITY Server CSI and SI—32 MB of flash Read only Memory (ROM) and 32MB of Dynamic Random Access Memory (DRAM).
    - Server R—4 TN1650B memory circuit packs for a total of 128 MB DRAM.
    - S8100/S8300/S8500 Media Servers—Up to 512 MB SDRAM. The S8700 is based on a commercial, Intel-based server platform running the Linux OS, and so can use as much memory as the server platform supports.
- **Storage**—All systems store translations in nonvolatile memory, either on a Personal Computer Memory Card International Association (PCMCIA) flash memory card, a hard disk drive, and/or an optical drive.
- **Input/Output (I/O) circuits**—These act as interfaces between the SPE and the TDM bus and packet bus.
- **Maintenance interface**—Connects the system to an administration terminal and monitors power failure, clock signals, and temperature sensors.

ACM Software Architecture

The ACM system consists of two main components, as illustrated on the ACM Software Architecture Diagram:

- The operation system:

    - The DEFINITY Server CSI, SI, and R systems run the Avaya proprietary Oryx/Pecos real-time, multiprocessing OS.

    - The remaining systems run either the Microsoft Windows 2000 Server or Redhat Linux OSs.

- Applications layer consisting of three major subsystems:

    - **Call processing**—Starts up and completes calls and manages voice and data in the system

    - **Maintenance**—Detects faults, recovers operations, and performs tests in the system

    - **System management**—Controls the internal processes necessary to install, administer, and maintain the system

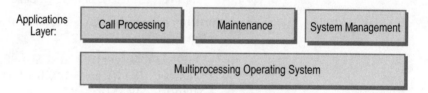

ACM Software Architecture

DEFINITY DCS Private Networks

The DEFINITY ECS supports both advanced Call Center integration and limited packet voice communications over TCP/IP data networks.

LAN Gateway

This configuration works with PC LAN applications that support the CallVisor Adjunct-Switch Application Interface (ASAI). ASAI is a DEFINITY-specific application programming interface (API) that provides commands and messages that enable features such as event notification and call control. Computer Telephony Integration (CTI) link applications that use ASAI can more easily access these services on a DEFINITY system.

For example, a call center application that uses ASAI could let agents access all job-related resources, such as the order-processing database, company World Wide Web (Web) site, telephone system, voice mail system, and fax machine, from a single interface on the PC.

IP Trunking

The DEFINITY ECS supports packet telephony over IP trunks. IP trunking is a good choice for basic corporate voice and fax communications, where cost is a major concern. IP telephony calls travel over an intranet or the Internet, rather than the public telephone network. Thus, for the most common types of internal corporate communications, IP trunks offer considerable savings.

IP trunking is usually not a good choice for applications where calls have to be routed to multiple destinations (as in most conferencing applications) or to a voice messaging system. This is because IP-trunk calls are compressed to save network bandwidth. Each cycle of compression and decompression results in a loss of data that degrades the final quality of the signal. This is not a problem in normal voice or fax calls, because they go through two or three compression cycles at most. However, multipoint conference calls and most voice messaging systems add too many compression cycles for acceptable quality. Companies that need true multipoint voice conferencing with packet telephony should consider using a data network architecture such as H.323.

DCS Private Networks

When two or more switches are connected via tie trunks, they form a private network. Two types of private networks can be formed using ASM servers:

- **Main-satellite/tributary (MS/T)**—A network of switches in which a main switch is fully functional, and provides attendants and CO trunks for connected satellite switches. Tributary switches are connected to the main switch and may have their own attendant and CO trunks. The main switch may be connected to one or more electronic tandem networks.

- **Electronic tandem network (ETN)**—A wide-area network of switches in which a call can tandem through one or more switches on its way from the originating switch to the destination switch. ETNs have a uniform numbering plan, automating alternate routing, and automatic route selection. An

ETN can be combined with a software defined network (SDN) to form a hybrid ETN/SDN network.

The switches in MS/T or ETN networks must be provisioned with special networking software packages.

When a call is transferred within a private network and improvements can be made in costs, the call's connection between the switches can be replaced with new connections while the call is active. This feature is known as Path Replacement with Path Retention and is implemented as DCS with Rerouting (Path Replacement) or Q.SIG Path Replacement with Path Retention or Stand-alone Path Replacement. Stand-alone Path Replacement is the process of rerouting an established call over a newer, more efficient path, after which the old call is "torn down," leaving those resources free. Path Replacement offers potential savings by routing calls more efficiently, thus saving resources and trunk usage. Additionally, transcoding or the conversion of a voice signal from analog to digital or digital to analog (with or without compression/decompression algorithms) is reduced, maintaining better voice quality between tandem switches.

DCS

Distributed communications system (DCS) is a messaging overlay for ETN or MS/T networks, that allows multiple switches to function as a single system. To enable DCS, switches require special DCS software in addition to the ETN or MS/T software.

DCS provides signaling connections between network nodes, so that certain call features can operate transparently across the network. By using DCS, these transparent features appear to operate as if all of the networked switches were actually one switch. For example, the DCS Call Coverage feature enables calls to an extension on one switch to be answered by extensions on a remote switch.

DCS signaling is done out-of-band; signaling messages are sent over a separate path from voice transmissions and the call-control data required to set up, maintain, and tear down a call connection. Thus, in addition to the normal tie trunk connections for voice and call-control data, DCS requires a special signaling connection between any two switches. This signaling link can be implemented in one of four ways, as shown on the DCS Connectivity Diagram:

- DCS over ISDN-PRI D-channel (DCS+)

- PPP over analog trunks

- ISDN D-channel signaling mapped to Avaya proprietary BX.25 protocol. The Processor Interface (PI) or Packet Gateway (PGATE) circuit pack is used to support existing BX.25 connections, though Avaya does not ship the PI interface in R7 and later systems. The Network Control/Packet Interface (NetPkt) circuit pack replaces the PI circuit pack. One DEFINITY Server acts as a gateway node to map ISDN D-channel data to the BX.25 protocol.

- TCP/IP packets carried over either a PPP connection or a 10/100Base-T Ethernet connection. A CLAN circuit pack supports DCS over PPP or Ethernet connections. DCS signaling for VoIP connections is sent over H.323 trunks through the CLAN, IP-Interface, IP Media Processor, or IP Server Interface (IPSI) circuit packs.

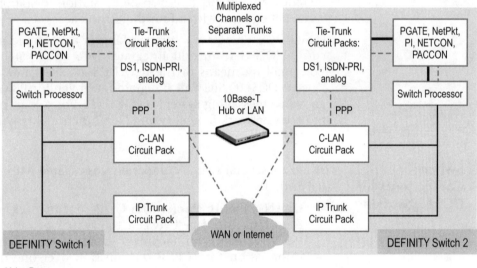

DCS Connectivity

For all connection types except IP Trunk, the call signaling and voice data are sent together over tie trunk facilities as TDM frames. For all connection types except CLAN Ethernet, the DCS signaling data is sent as packets over a permanent virtual circuit (PVC) on tie-trunk facilities.

For IP Trunk connections (R7 IP Processor operating in IP trunk mode or H.323 trunk), the voice data is sent over TCP/IP facilities as IP packets; each packet can potentially take a different route through the network. The DCS signaling can be sent over tie trunk PVCs or TCP/IP facilities (R7) or over TCP/IP facilities (DCS+ over H.323 trunks).

When a DCS network uses two or three different DCS signaling types, one or more switches must act as a gateway. The gateway is connected between two switches that use different signaling protocols; it enables communication by converting the signaling messages from one protocol to another. A gateway can convert between two, or all three, of the signaling protocols, but only one protocol can be used for DCS signaling between any two switches.

DCS, used for Voice Mail Interworking, is limited to a four-digit or five-digit dialing plan between network subscriber stations and voice mail. DCS is also limited to a three-digit trunk access code. By comparison, Q.SIG, used in Avaya's Mulitvantage software, includes an extended, seven-digit, station-to-station, network-wide dialing plan with a four-digit trunk access code.

Q.SIG/DCS Voice Mail Interworking is an enhancement to the current Q.SIG feature. It integrates DCS and Q.SIG Centralized Voicemail by means of the new DCS+/Q.SIG gateway. Switches labeled DCS+/Q.SIG will integrate multi-vendor PBXs into a single voice messaging system. Q.SIG/DCS Voice Mail Interworking provides network flexibility and DCS functionality without a dedicated T1.

Switch Components for DCS Networking

There are several switch components associated with DCS networking:

- **CLAN Circuit Pack**—The CLAN circuit pack provides the Data Link interface between the switch processor and LAN/WAN transmission facilities. CLAN prepares the signaling information for TCP/IP transmission over one of two pathways: an Ethernet LAN, or a PPP connection. For an Ethernet connection, the signaling data is sent out on a 10Base-T Ethernet network that is connected directly to the CLAN Ethernet port. For a PPP connection, CLAN inserts the signaling data on a TDM channel, alongside the DS1 bit streams that carry voice transmissions.

- **IP Interface Circuit Pack**—The IP Interface circuit pack enables two switches to transmit voice and DCS data (R7 IP trunk mode) between them over an IP network. The DCS signaling data, on a separate path from the voice data, can use either TCP/IP or X.25 protocols.

- **Tie Trunk Circuit Pack**—Tie trunk circuit packs allow the switch to connect to various types of physical transmission facilities, such as analog trunks, DS1 channels, or ISDN-PRI.

- **IP Media Processor**—The IP Media Processor replaces the IP Interface circuit pack in R9 and later servers. It supports H.323 trunks.

- **Interface Between Tie Trunks and Processor**—Depending on the DEFINITY Server in use, various circuit packs form the interface between the tie trunk circuit pack and the switch processor:

  - The Network Controller (NETCON) board is used in older (pre-R7) systems; it connects the process and port circuit packs to the TDM bus.

  - The Packet Controller (PACCON) board is used in older (pre-R7) systems; it provides an interface to the processor for D-channel signaling over the packet bus.

  - The Network control/Packet Interface (NetPkt) board replaces both the NETCON and PACCON boards in newer (R7 and later) systems.

  - The Processor Interface (PI) board is used to terminate X.25 connections in older (pre-R7) systems.

  - The Processor Gateway (PGATE) board is used in Server R systems; it connects the processor to the packet bus and terminates X.25 signaling.

IP-Enabling DEFINITY Servers

Avaya is moving from the traditional PBX market to the IP PBX market. Where the DEFINITY line (prior to R9) supported IP trunking in a limited capacity, the latest IP-enabled PBXs, media servers, and gateways support full-blown IP-telephony.

A customer can update an existing DEFINITY Server to an IP-enabled PBX with an upgrade package purchased from Avaya or a business partner. Beginning with R9, a DEFINITY Server can be updated to support H.323 and IP trunks and IP soft- and hardphones.

Besides the requisite software upgrade, two circuit packs are required to be installed in the DEFINITY server to support IP telephony:

- The TN2302 IP Media Processor circuit pack for H.323 voice processing

- The TN799 CLAN for signaling

The TN802 IP Interface can be installed to support R7 IP trunk mode, if needed.

IP Media Processor

The TN2302AP IP Media Processor circuit pack supports the ITU-T H.323v2 recommended standard for transmitting packetized multimedia traffic. It performs echo cancellation, silence suppression, jitter buffering, DTMF detection, and audio encoding and decoding using the G.711, G.723, or G.729A/B/AB codecs. It provides IP connectivity between DEFINITY servers and Avaya Media Servers, and supports H.323 trunk groups configured as tie trunks to carry DCS+, Q.SIG, and DID traffic, as well as enabling connectivity to other vendors' H.323v2 switches.

The IP Media Processor supports hairpinning and shuffling, two features designed to reduce the use of server TDM resources and improve performance and quality on IP-endpoint-to-IP-endpoint calls. Hairpinning routes audio channels from one IP endpoint through the Media Processor circuit pack to the other IP endpoint without the use of the TDM bus. Shuffling reroutes an audio channel directly between IP endpoints once TDM resources are no longer needed (the voice no longer passes through the IP Media Processor circuit pack).

CLAN

As mentioned earlier, the TN799 CLAN circuit pack supports signaling over IP networks. In an IP-enabled server, the CLAN circuit pack handles call control for all IP endpoints connected to the server, while the IP Media Processor handles audio processing. The CLAN can also support communications between DEFINITY servers and adjuncts, such as a CMS, voice mail system, and call detail report (CDR) systems. The CLAN circuit pack supports up to 500 simultaneous UDP connections.

Upgrading DEFINITY Servers to S8700 Multi-Connect

DEFINITY servers can be upgraded to S8700 Multi-Connect systems running ACM. This is a three stage process where:

1. An Avaya S8700 Media Server replaces the G3 processor

2. TN2302AP and TN799DP circuit packs are installed in the PNs

3. Multiple systems are consolidated to simplify administration

The S8700 Media Server uses the Linux OS on an Intel processor-based server platform to enable full IP support in DEFINITY servers running ACM. The S8700 is discussed in more detail later in the module.

The order of the first two upgrade process stages is unimportant, which means an existing IP-enabled system can be upgraded with the addition of the S8700 server.

Stage 1: Processor Replacement

When an S8700 server replaces the G3 processor, the means with which the processor and PNs communicate changes. In order to allow PNs to be directly controlled by the S8700, TN2312 IP Server Interface (IPSI) circuit packs must replace the PN call control components. If a PN is not upgraded, the S8700 controls it via an IPSI-connected PN connected, in turn, to a CSS or the ATM network. The S8700 Multi-Connect Diagram illustrates this configuration.

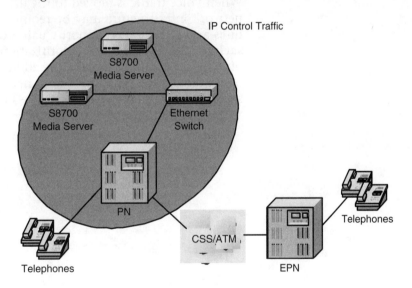

S8700 Multi-Connect

Stage 2: IP Enable the PNs

A PN is IP-enabled by installing the CLAN and IP Media Processor circuit packs. It then becomes a media gateway, capable of integrating IP and TDM devices, allowing non-IP and IP devices access to all ACM features.

Stage 3: Server Consolidation

S8700 servers enable enterprises to consolidate remote DEFINITY systems into a single, centrally administered entity. For survivability, remote sites can host local survivable processors (LSPs) to take over local call processing if they lose communications with the S8700. Each site has an S8300 media server and G700 media gateway installed. Licensing files determine how the S8300 operates both in normal circumstances and in the case of a network failure.

If a media gateway loses contact with its primary server, it will first try to reestablish communications. If communication still fails, it will then try to contact the LSP. After it locates a functional LSP, the media gateway reboots and falls under the call control of the LSP. The LSP provides full call processing capabilities while the S8700 is offline, and supports both non-IP and IP endpoints.

Upgrade the Data Network

When voice traffic is moved to the data network, upgrades to the network infrastructure may be required. Switches should replace hubs and should also support Quality of Service (QoS) protocols, such as VLANs and traffic prioritization. The network addressing plan must be able to support the additional IP devices. UPSs may be required, and WAN links may need additional bandwidth to support voice traffic between remote sites.

Activities

1. Match the following terms with the correct description:

   IP Trunking

   LAN Gateway

   Expansion Port Network (EPN)

   Center Stage Switch (CSS)

   Switch Node (SN)

   Processor Port Network (PPN)

   Distributed Communications System (DCS)

   a. Out-of-band signaling connecting ETN nodes

   b. Connects PC LAN applications to a DEFINITY switch

   c. digital networking component that contains the Switch Processing Element (SPE)

   d. Central interface between the PPN and EPNs

   e. Contains additional ports that increase the number of connections to trunks and lines

   f. Reduces the amount of interconnecting cabling between a PPN and EPNs

   g. Supports packet telephony over IP trunks.

2. Q.SIG extends the dialing plan between network subscriber stations and voice mail _____.

   a. From 4 or 5 digits to 7 digits

   b. From 5 digits to 8 digits

   c. From 3 digits to 8 digits

   d. Q.SIG does not affect the dialing plan in this way

3. Path replacement _____.

   a. Reduces transcoding

   b. Uses the most efficient path available

   c. Improves voice quality between tandem switches

   d. All of the above

4. The newest DEFINITY servers have _____.

   a. Full-blown IP telephony capabilities

   b. Limited IP telephony capabilities

   c. No IP telephony capabilities

   d. Limited IP trunking capabilities

5. What does hairpinning do?

   a. It reroutes audio channels directly between IP endpoints.

   b. It routes audio channels through the Media Processor circuit pack to the other IP endpoints without using the TDM bus.

   c. It performs audio encoding and decoding.

   d. It provides jitter buffering.

6. What does a CLAN circuit pack do?

   a. It supports signaling over IP networks.

   b. It handles call control for all IP endpoints connected to the server.

   c. It handles audio processing.

   d. Answers a and b are correct.

7. What does the NetPkt Interface do?

   a. It connects the process and port circuit packs to the TDM bus and provides an interface to the processor for D-channel signaling over the packet bus.

   b. It connects the processor to the packet bus and terminates X.25 signaling.

   c. It terminates X.25 connections in older systems.

   d. It allows the switch to connect to various types of physical transmission facilities, such as analog trunks, DS1 channels, or ISDN-PRI.

Extended Activities

1. Research the following two connectivity options for a PBX and list which vendors support each function and how each is implemented:

    a. LAN gateway

    b. IP trunking

Lesson 4—DEFINITY Server Hardware

This lesson presents a general overview of the equipment used in DEFINITY Server implementations. For more specific information, refer to the product documentation for each component.

Objectives

At the end of this lesson you will be able to:

- Name and describe the three types of DEFINITY system cabinets

- Plan the layout of an equipment room for a particular DEFINITY system

- Explain which environmental conditions must be controlled in an equipment room and why

 **Key Point**

All three DEFINITY system cabinets operate in basically the same way.

System Cabinets

System cabinets house all components, including the direct-current power supply (power converter). In particular, cabinets house the circuit packs that connect trunks and users to the core PBX circuitry. Trunk cards (or CO cards) allow trunks to be connected to the PBX; cards are specialized according to the type of trunk. Station cards connect the PBX to individual desk telephones. There are three types of cabinets:

- Compact modular cabinets (CMC1s)

- Single-carrier cabinets (SCC1s)

- Multicarrier cabinets (MCC1s)

Carriers

A cabinet contains at least one carrier, which holds circuit packs and connects them to power, the TDM bus, and the packet bus. A carrier is an enclosed shelf with vertical slots; circuit packs fit into connectors attached to the rear of the slots. The hardware capacity of a PBX depends on the total number of slots within each cabinet. There are five types of carriers:

- Control carrier (PPN cabinet only)
- Optional duplicated control carrier (PPN cabinet only)
- Optional port carrier (PPN and/or EPN cabinets)
- Optional expansion control carrier (EPN cabinets only)
- Optional SN carrier (PPN and/or EPN cabinets)

CMC1 Media Gateway

A Compact Modular Cabinet (CMC1) Media Gateway is only used as a PPN and provides standard reliability only (no components are duplicated for redundancy). It mounts on a wall (preferred) or sits on the floor (with a floor panel), as illustrated on the CMC1 Media Gateway Diagram.

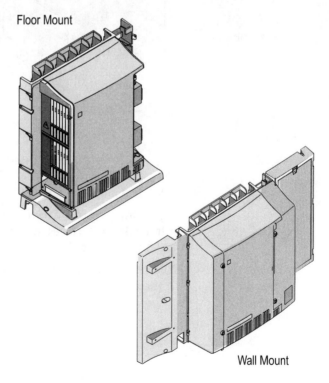

Floor Mount

Wall Mount

CMC1 Media Gateway

This cabinet has a single carrier, which contains universal port slots. Slot 1 is left vacant while the IPSI card resides in slot 2. The remaining eight slots house port and service circuit packs.

SCCl Media Gateway

Up to four Single-Carrier Cabinets (SCC1) Media Gateways can be stacked to form a single PN. DEFINITY Server SI supports up to three stacks of four cabinets each (three PNs), while the DEFINITY Server R supports up to 44 PNs. Avaya S8700 Multi-Connect supports up to 64 PNs. A typical SSC1 Media Gateway is shown in the SCC1 Media Gateway Diagram.

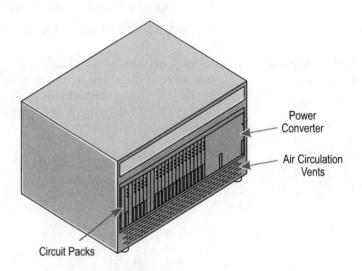

Power Converter

Air Circulation Vents

Circuit Packs

SCC1 Media Gateway

SCC1 Media Gateways are available in two configurations:

- Expansion control cabinet that contains additional port circuit packs, interfaces to the PPN, a maintenance interface, and a power supply

- Port cabinet that contains ports and a port supply to interface to the expansion control cabinet

MCC1 Media Gateway

As shown on the following diagram, a Multicarrier Cabinet (MCC1) Media Gateway is a 70-inch (178-centimeter [cm]) cabinet that houses up to five carriers, labeled A through E. The three types of MCC1 Media Gateways include:

- A port network containing port, IPSI, service, tone clock, and expansion interface circuit packs

- An SN carrier containing the CSS SNI circuit packs

- An expansion control carrier containing additional ports, interfaces to the PPN and other EPN cabinets, the maintenance interface, and optional interfaces to other EPN cabinets and/or an SN (in an ATM- or CSS-connected system)

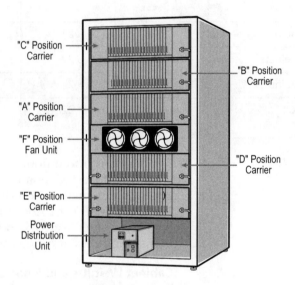

MCC1 Media Gateway

Floor Area and Load

Each DEFINITY Server cabinet has its own requirements for floor space and strength.

Floor Area

Floor area requirements vary between cabinets. For maintenance access, floorplans typically allocate space around the front, ends, and rear of the cabinets. Dimensions and clearances for typical cabinet configurations are listed in the DEFINITY Server Cabinet Configurations Table.

DEFINITY Server Cabinet Configurations

Cabinet Type	Height (inches)	Width (inches)	Depth (inches)	Clearance (inches)
CMC1 1 cabinet	25.5	24.5	12	Left, Right and Front: 12
2 cabinets	51	24.5	12	
3 cabinets	76.5	24.5	12	
SCC1 1 cabinet	20	27	22	Between cabinet and wall: 38
2 cabinet	39	27	22	
3 cabinets	58	27	22	
4 cabinets	77	27	22	
MCC1	70	32	28	Rear: 38 Front: 36

Floor Load

The equipment room floor must meet the commercial floor loading code of at least 50 pounds (lbs) per square foot (242 kilograms [kg] per square meter). Additional equipment room floor support may be required if the floor load is greater than 50 lbs per square foot. The Cabinet Weights and Floor Loading Requirements Table presents examples of cabinet weights and floor loadings.

Cabinet Weights and Floor Loading Requirements

Cabinet Type	Weight (pounds)	Floor Loading (lbs per sq. ft.)	Remarks
CMC1	58		Typically wall mounted
SCC1	125	31	
MCC1	200 to 800	130	Includes auxiliary, global AC, and global DC cabinets

Floorplan Guidelines

DEFINITY Server floorplans vary with the size and shape of the equipment room and the extent of future growth. Future growth includes a new or upgraded system, adjuncts and peripherals, and the main distribution frame (MDF). Also, in the United States, placement of equipment is governed by the Americans with Disabilities Act (ADA), which requires enough working room for employees who use wheelchairs or other adaptive equipment.

For floor-standing cabinets, reserve the area behind a cabinet for the MDF and cable slack manager. For wall-mounted cabinets, reserve the area beside the cabinets for the MDF.

CMC1 Configuration Guidelines

The MDF (cross-connect) is either to the rear or right of the cabinet. To allow service access, the table for the management terminal and optional printer should be located away from the equipment area. In an installation where no MDF is present, an MDF can be installed in the cabinet right panel. A typical setup for this type of cabinet is illustrated on the Typical CMC1 Floorplan Diagram.

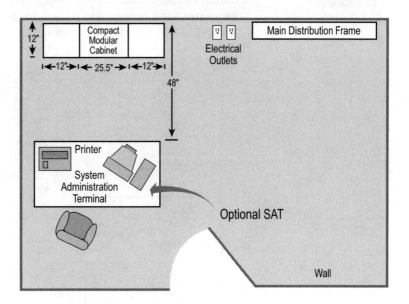

Typical CMC1 Floorplan

Before installing a CMC1:

- Locate the power outlets outside the MDF area. The outlets must not be controlled by a wall switch or shared with other equipment. Other devices may connect to other outlets on the same circuit.

- Locate the trunk/auxiliary field inside the MDF, if desired.

- Ground the system. In most cases, all that is required is to plug the system into a properly grounded (three-prong) receptacle.

- Install earthquake protection, if required.

- Use the appropriate receptacle or cord set. Each cabinet requires either a National Electrical Manufacturers Association (NEMA) 5-15R receptacle (or equivalent) for United States installations, or a local International Electrotechnical Commission (IEC) 320 cord set (or equivalent) for non-United States installations.

SCC1 Configuration Guidelines

In an SCC1 configuration, the MDF can be directly behind the cable slack manager. To allow service access, the table for the management terminal and optional printer should be located away from the equipment area. This type of setup is illustrated on the Typical SCC1 Floorplan Diagram.

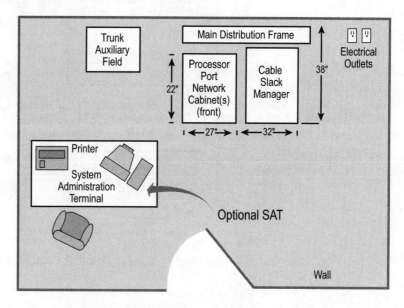

Typical SCC1 Floorplan

Before installing an SCC1:

- Locate the power outlets outside the MDF area. The outlets must not be controlled by a wall switch or shared with other equipment. Other devices may connect to other outlets on the same circuit.

- Locate the trunk/auxiliary field inside the MDF, if desired.

- Ground the system. In most cases, all that is required is to plug the system into a properly grounded (three-prong) receptacle.

- For fiber connections between PNs, use a 20-foot (ft) (6.1-meter [m]) multimode fiber optic cable in the standard core/cladding size of 62.5/125 micrometers (μm).

- Install earthquake protection, if required.

- Use the appropriate receptacle or cord set. Each cabinet requires either a NEMA 5-15R or NEMA 5-20R receptacle (or equivalent) for United States installations, or a local cord set (or equivalent) for non-United States installations.

MCC1 Configuration Guidelines

In an MCC1 configuration, the MDF is directly behind the cable slack manager. To allow service access, the table for the management terminal and optional printer should be located away from the equipment area. This type of setup is illustrated on the MCC1 Floorplan Diagram.

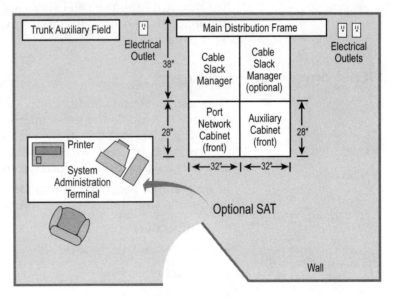

MCC1 Floorplan

Before installing an MCC1:

- Locate the power outlets outside the MDF area. The outlets must not be controlled by a wall switch or shared with other equipment. Other devices may connect to other outlets on the same circuit.

- Locate the trunk/auxiliary field inside the MDF, if desired.

- Ground the system. In most cases, all that is required is to plug the system into a properly grounded (three-prong) receptacle.

- For fiber connections between PNs, use a 20-ft (6.1-m) multi-mode fiber optic cable in the standard core/cladding size of 62.5/125 μm.

- Install earthquake protection, if required.

- For PPN cabinets, use either a NEMA 5-50R receptacle (or equivalent) or NEMA L14-30R receptacle (or equivalent) power outlet, or 220V AC, 50 to 60-hertz (Hz) power outlet for a global AC cabinet.

- For an auxiliary cabinet, use a NEMA 5-20R receptacle (or equivalent).

- Allow at least 3 ft (91.4 cm) clearance in front of the cabinet to open the door.

MDF
The Main Distribution Frame (MDF) equipment is located a specified distance from the DEFINITY cabinets and must meet specific requirements. An optional MDF can be installed in the CMC1 right panel.

Environmental Conditions

The environmental conditions in an equipment room can either enhance or degrade the performance of electronic devices. When installing a DEFINITY system, the following environmental conditions must be kept within specified ranges. Conditions that exceed these limits may reduce system life or affect its operation.

In general, the performance of a DEFINITY installation is affected by the following environmental conditions in the equipment room:

- Temperature
- Altitude and air pressure
- Humidity
- Air purity

- Lighting
- Radio frequency noise

Temperature

Temperature affects the performance of electronic components. As the temperature rises, resistance and capacitance also rise; as the temperature falls, resistance and capacitance decrease. Thus, if electronic systems become too hot or too cold, they may malfunction.

Fortunately, the DEFINITY system functions well at the same ambient temperature conditions that are comfortable for people.

- Maximum performance is achieved at an ambient room temperature between 40 and 120 degrees Fahrenheit (°F) (4 and 49 degrees Celsius [°C]).

- For continuous operation, the maximum temperature should not exceed 110° F (43° C).

- For short-term operation (not more than 72 consecutive hours or 15 days in a year), the maximum temperature should not exceed 120° F (49° C).

Therefore, it is vital to keep an equipment room well ventilated. Ventilation is important because the equipment itself generates a significant amount of heat while operating. Typical heat dissipation figures for compact modular, single-carrier, and MCC1s are shown in the Heat Dissipation Table.

Heat Dissipation

Cabinet Type	Number in Stack	With Terminals?	BTUs/Hour
CMC1	1	No Yes	202 378
SCC1	1 (4 max)	Yes Yes	438 1,436
MCC1	1	No Yes	1,058 1,662

Altitude and Air Pressure

Air pressure does not directly affect the performance of the DEFINITY. However, it strongly influences other factors, such as temperature and humidity, that do affect system performance.

At altitudes above 5,000 ft (1,525 m), the maximum short-term temperature limit drops by 1°F for each 1,000 ft (305 m) of elevation above 5,000 ft. For example, at sea level, the maximum short-term temperature limit is 120°F (49°C). At 10,000 ft (3,050 m), the maximum short-term temperature limit is 115°F (46°C).

DEFINITY's normal operating air pressure range is summarized in the DEFINITY Air Pressure Range Table:

DEFINITY Air Pressure Range

Variable	Minimum	Maximum
Air pressure	9.4 pounds per square inch or 648 millibars	15.2 pounds per square inch or 1,048 millibars
Altitude (relative to sea level)	-820 feet (-250 meters)	11,500 feet (3,500 meters)

Humidity

Humidity and temperature are closely related, because water vapor in the air can store a great deal of heat. In humid conditions, more of the heat generated by electronic equipment tends to be held by the air.

The acceptable humidity of an equipment area depends on the ambient temperature. Therefore, humidity and temperature must always be considered together.

- At ambient temperatures up to 84°F (29°C), acceptable relative humidity can range from 10 to 95%.

- As temperature rises above this level, the maximum relative humidity gradually decreases. At 120°F (49°C), the maximum relative humidity is only 32%.

- The recommended temperature and humidity range is 65° to 85°F (18° to 29°C) at 20 to 60% relative humidity.

The main point to remember is that warm air holds more moisture than cool air. Lowering the air temperature will reduce humidity by reducing the ability of the air to hold water. Therefore, by air conditioning an equipment room, you can correct both high temperature and high humidity.

Air Purity

Impurities in the air can damage an electronic system in two ways:

- Conductive contaminants, such as metallic dust, may cause short circuits.

- Dust can clog a cooling system and make the equipment run too hot.

Given these concerns, all DEFINITY cabinets contain an air filter to reduce the amount of particulates flowing through the equipment. Do not install a cabinet where the air may be contaminated by excessive dust, lint, carbon particles, paper fiber contaminants, or metallic contaminants. Replace or clean the filter as indicated in the system manual.

Lighting

Although DEFINITY does not need light, the administrative staff does. Occupational Safety and Health Administration (OSHA) standards recommend a light intensity of 50 to 70 footcandles (538 to 753 lumens per square meter) as a level suitable for detailed work.

Radio Frequency Noise

Electromagnetic fields near system control equipment may introduce noise into the system through trunk cables, station cables, or both. Place the system and cable runs in areas where high electromagnetic field strengths do not exist.

Radio transmitters (AM or FM), television transmitters, induction heaters, motors with commutators of 0.25 horsepower (187 watts) or greater, and similar equipment are leading causes of interference. Small tools with universal motors are generally not a problem when they operate on separate power circuits. Motors without commutators generally do not cause interference. Field strengths below 1.0 V per meter are unlikely to cause interference.

Measure weak fields with a tunable meter. Measure field strengths greater than 1.0 V per meter with a broadband meter.

Estimate field strengths of radio transmitters by dividing the square root of the emitted power (in kilowatts) by the distance from the antenna (in kilometers). This yields the approximate field strength in volts per meter and is relatively accurate for distances greater than approximately 0.5 wavelength (150 m for a frequency of 1,000 kilohertz [kHz]).

Cabinet Power Requirements

Each DEFINITY cabinet also has particular AC- and DC-power source requirements. The design and implementation of any new system must comply with the National Electrical Code (NEC), as well as any applicable local building or electrical codes.

AC Power

Power feeders from a dedicated AC-power source (usually the public power system outside the building) connect to an AC-load center. The AC-load center distributes the power to many separate circuits throughout the building. Each one of those circuits dstributes power to one or more receptacles (outlets).

The power cord, from the AC-power distribution unit in each MCC1, and AC-power supply in each SCC1, plugs into a receptacle. The outlet that serves a DEFINITY cabinet must not be controlled by a wall switch or shared with other equipment.

Either of the following power sources can supply 60-Hz AC power to R7 and later systems:

- Single-phase, 4-wire, 120/240V AC supplying 240V AC

- Three-phase, 5 wire, 120/208V AC supplying 208V AC

(The differences between single-phase and three-phase electrical power are beyond the scope of this course.)

Note: If you are ever unsure about the nature of a power source, consult an electrician before attempting to plug an electronic device into it.

DC Power

DC power is simpler and does not require power conditioning. DC-powered cabinets containing a J58890CF power distribution unit require a -42.5 to -56V DC source, at up to 75 amperes (amps). A DC-powered SCC1 uses a plug-in DC power supply in the power supply slots.

Power Conditioning

Many electric utilities are only obligated to deliver electricity at 120V, plus or minus 10 percent. And, electrical power is vulnerable to many types of electromagnetic interference/radio frequency interference (EMI/RFI) (noise) on its way through the power distribution system. While this "dirty" power is fine for lights and stoves, it is not consistent enough for delicate electronic devices.

Power conditioning is a method of protecting electronic components, such as PBXs, from fluctuations in the power source. Power conditioning provides "clean" power by preventing the following problems:

- **Transients**—Short, extreme spikes or surges in voltage.

- **Noise or static**—Smaller fluctuations in voltage. In the DEFINITY, EMI filters suppress noise voltage on the AC input line to the unit.

- **Brownouts**—Temporary reduction in electrical power level.

- **Blackouts**—Temporary failure of electrical power.

The most commonly used component for power conditioning is an online UPS system. An online UPS operates continuously, and actively regulates power quality as electricity passes through it. Thus, it can protect equipment from power problems, as well as provide power in the case of power failure. In contrast, an offline UPS is only activated when the power fails; it can protect equipment against blackouts, but cannot provide true power conditioning.

Cabinet Power Components

A typical AC-power distribution unit for an MCC1 contains circuit breakers or fuses, a ring generator, optional batteries, and an optional battery charger. The power distribution cables carry 120V AC during normal operation and 144V DC from optional batteries if AC power fails. Another cable connects 120V AC to the battery charger.

DC Power Relay

A DC power relay disconnects the batteries from a system when using AC power. The relay also disconnects the batteries if power fails for more than 10 minutes in a standard reliability system, 5 minutes in high and critical reliability systems, and 10 minutes in an EPN. This protects the batteries from over discharging.

Ring Generator

A ring generator converts the -48V DC input to a 67V AC to 100V AC, 20 or 25 Hz ringing voltage. The analog line circuit packs use this AC voltage output to ring voice terminals. The AC outputs are routed from the ring generator to port carriers, expansion control carriers, and control carriers.

Fuses

Fuses protect a system from an excessive volume of current flow (amperage). Twenty-amp fuses protect the power on each cable going from the AC-power distribution unit to power converters in the carriers.

Power Backup

If AC power fails, three 48V DC batteries power the system for 10 seconds in a PPN cabinet, 15 seconds in an EPN cabinet, and 10 minutes in the control carrier in a standard reliability system. The batteries also supply system power for five minutes in the control carrier in high and critical reliability systems, and 10 minutes in the expansion control carrier in the A position of an EPN cabinet (Server R only).

UPS

An external uninterruptible power supply (UPS) provides a longer backup time than holdover batteries. The unit connects from the AC power source to a cabinet's AC-power cord. If AC power fails, the UPS unit supplies its own AC power to the cabinet. Holdover time is the time the batteries power the equipment in the absence of line power. UPS holdover times vary from less than 10 minutes to up to eight hours, and can replace the batteries and battery charger. In a high-reliability environment, it is important to choose a UPS that provides adequate holdover time.

Conventional digital and analog telephone sets are powered by the telephone line itself. Because of this, when the PBX goes on backup power, so do the telephones. However, this is not the case with IP telephones; power cannot be packetized. Each telephone set receives power from its own local power supply or over spare wires in the Ethernet cable supplied by either a Power over Ethernet (POE)-capable inline power supply or Ethernet switch. The IEEE 802.3af standard specifies how network devices may be powered over the Ethernet network. Regardless of how the phones receive power, if high reliability is required, their power sources must have UPSs installed as well.

Another consideration when using a UPS for backup power is the space they take in equipment rooms, equipment closets, and work areas. A UPS may be as small as a power strip or as large as several equipment racks. When planning a DEFINITY or any other voice system installation, you must include in the plan room to place the backup power systems.

Activities

1. If possible, try to do one of the following:

   a. Go on a field trip to a LEC to see switching gear such as that described in this lesson.

   b. Go to your school or business and see the telephone closet that contains networking and telephony gear. Locate the telephone switching equipment and discuss the components of the system.

2. What is holdover time?

   a. The duration that batteries power equipment in the absence of line power

   b. The amount of time a power line spike affects the equipment

   c. The amount of time a buffer retains data after loss of power

   d. All of the above

3. From where does an IP phone get its power?

   a. Its own local power supply

   b. From the phone line

   c. Power over Ethernet lines

   d. Answers a and c are both feasible

   e. All of the above

Extended Activities

1. Using the Web, research information on the DEFINITY Server system. Try to determine the cost per circuit for the following three cabinet types, fully loaded:

   a. CMC1

   b. SCC1

   c. MCC1

Lesson 5—IP Telephony

Traditional PBX environments have a long-standing history of performance and have established high expectations for performance. New technologies have made it possible to converge networks into a single cable infrastructure, a single protocol, and a single packetized delivery method. IP Telephony and Voice over IP (VoIP) products have emerged to meet the opportunities and challenges in this growing arena. New standards-based, open architecture systems are available to perform call processing, protocol conversion, and provide extensive feature sets, all in an IP network environment. Media servers, gateways, and combination voice and data switches are providing businesses big and small with the tools they need to build, scale, and take their networking into the future.

Communication servers that seemed "bleeding edge" technology are now available and in high demand. They are proving their ability to service toll-quality, voice-grade communications of 99.999 percent reliability. They are proving their ease of administration, integration, and opportunities for applications that were difficult, if not impossible, to implement in the past. Finally, they are proving a high return on investment for companies that embrace the change from traditional time division multiplexing (TDM) to IP communications.

Objectives

At the end of this lesson you will be able to:

- Identify the features and benefits of Avaya Communication Manager (ACM) software
- Identify the protocols that ACM supports
- Describe the functionality of a Media Server
- Describe the characteristics of Avaya's S8700, S8500, S8300, and S8100 Media Servers
- Describe the functionality of a Media Gateway
- Describe the characteristics of Avaya's G350, G600, G650, and G700 Media Gateways
- Describe the functionality of Avaya's IP Office
- Identify the telephones supported by Avaya's Media Servers

**Key
Point**

Avaya Communication Manager is the call-processing and feature-rich software that provides IP communication capabilities across a diverse product line.

Avaya Communication Manager

This software component is what makes Voice over IP and IP telephony capabilities, as well as more traditional communication functions, possible on Avaya's diverse product line of DEFINITY servers, media servers, and media gateways. Communication Manager is the software responsible for providing functions as simple as call completion, while at the same time offering very advanced processing functions for huge contact centers and multinational organizations. This software can run on Microsoft Windows 2000, Linux, or Avaya's DEFINITY UNIX. Avaya Communication Manager takes media servers and gateways into the future with various IP communication options, converged networking capabilities, and call-processing power that is three times that of the biggest DEFINITY server.

In addition to call processing, the software is responsible for the hundreds of time-saving and business-critical features necessary from today's phone systems. It also offers a wide range of expansion and networking options to handle everything from international call centers to allowing a single person to answer an office phone from home or while on the road. This multiprotocol software supports industry-standard application program interfaces (APIs) to provide a huge range of add-on software options and third-party adjuncts, services, or features. The options are virtually endless.

Avaya Communication Manager is also responsible for the stable and reliable environment necessary in voice communications. It maintains 99.999% availability, which is achieved through well-developed code, internal monitoring, and diagnostics capabilities. An Integrated Management component is included with the software to provide administration and monitoring capabilities in a converged network environment. It provides extensive scalability for both traditional TDM environments and IP. To prevent outages, it even offers advanced routing functionality that can detect power outages, network quality issues, or low network bandwidth and fail-over to an alternate trunk.

Avaya Communication Manager supports the following protocols:

- **H.323**—This protocol is used along with associated protocols for packetized multimedia and IP communications

- **Session Initiation Protocol (SIP)**—Used for IP station and trunk communications in future releases

- **Distributed Communication Systems (DCS and DCS+)**—Used for connection to Avaya DEFINITY servers

- **Q.SIG**—The industry standard protocol for ISDN communications and connectivity between different vendors' equipment

- **Diffserv**—A code point used to prioritize traffic for Layer 3 through the network

- **Reservation Protocol (RSVP)**—Used to shape traffic and allocate bandwidth

- **802.1p**—The industry standard protocol for Layer 2 prioritization

- **802.1Q**—The industry standard protocol for Layer 2 virtual local area network (VLAN) tagging

For computer telephony integration (CTI), as well as other add-on software features and services, Avaya Communication Manager supports the following API standards:

- **Telephony Server API (TSAPI)**—Developed by Novell and AT&T for server platforms

- **Telephony API (TAPI)**—Developed by Microsoft and Intel for both client and server platforms

- **JAVA Telephony API (JTAPI)**—Used for Web-based telephony applications

- **Directory API (DAPI)**—Used in conjunction with light weight directory access protocol (LDAP), which is used for lookup services

- **Adjunct Switch Application Interface (ASAI)**—Used on the DEFINITY for voice response or messaging adjuncts

Media Servers

A media server is the newer terminology for describing the functionality traditionally provided by a PBX. Other names for a media server include communication server, call server, call controller, media gateway controller, and gatekeeper. This device is often simply called a switch, although its functionality far exceeds that nomenclature today. Media servers are used to process calls, make routing decisions, provide connectivity, and allow access to a wide range of features and functions, as well as adjuncts and other software components. When more than one network type or protocol is used, a media gateway is also required. Media gateway functionality will be discussed in the next section. For now, we will look into the various media server products Avaya offers.

S8700 Media Server

The S8700 is an enterprise class media server, which delivers the highest level of service for large and complex network environments. It is based on the Linux operating system and utilizes Intel processor(s). It is designed for scalability by offering connectivity for 12,000 IP phones and up to 36,000 total IP, digital, and analog stations. In addition, the S8700 offers the highest connectivity capability with 8,000 trunk connections. Supported trunk types include: analog, T1/E1, ATM, ISDN BRI/PRI, and IP trunks using either H.323 or SIP. Networking options increase even more with the ability to connect up to 250 media gateways (G350, G650, G700), and up to five expansion port networks, meaning remote offices and major branch locations can be distributed while maintaining a high level of service. This equates to over a million endpoints being connected through an enterprise network. The S8700 can also be configured for the SCC1 and MCC1 to leverage and expand existing phone equipment.

The processing capabilities of this media server are equally impressive, guaranteeing 300,000 business hour call completions. The S8700 is also designed for exceedingly high availability and reliability to maintain toll quality performance expected in a server of this caliber. It builds on a two-server construct, which allows it to act as an active server when performing under optimal conditions and a stand-by or backup server, which takes over in the event that the first server experiences issues. Servers can be connected up to six miles apart using single mode fiber.

**Availability/
Reliability**

Voice and other real-time applications found in today's networks are not tolerant of downtime. An outage that would have been problematic in a data network is now considered catastrophic when both voice and data services are impacted. Convergence therefore requires data networks to "raise the bar," so to speak, and measure up to the five 9s (99.999%) of availability demanded from a toll-quality network environment.

Avaya deals with availability on two different levels and has additional objectives, which help in the network design process. First and foremost, the hardware and software components are designed to minimize failures. All products are built on this principle. To help reduce the impact of failures when they occur, the other objective is to reduce downtime significantly by using advanced error detection and correction mechanisms. The objectives for availability include:

- Failures of each component and sub-system must be infrequent. This is measured as the Mean Time Between Failures (MTBF).

- System outages must be infrequent. This is measured as the Mean Time Between Outages (MTBO).

- When there is a failure or outage, the impact must be minimized and isolated; recovery must be speedy. This is measured as Mean Time to Recovery (MTTR).

- The system can collect its own performance statistics. Although this is not measured with a specific component, it is an important part of the overall diagnostics process.

The S8700 is offered with three different levels of availability/reliability, depending upon the needs of the business. This is based on the hardware and software components and cannot account for power failure or telecommunications provider outages. Obviously, power backup components, such as UPSs are a minimum requirement to help reduce the impact of a power outage. A generator may be necessary in more critical availability situations. The three availability levels are duplex, high reliability, and critical reliability server configurations, which have the following characteristics:

- **Duplex (standard)**—This server configuration is designed for normal business operations that can tolerate about four hours of downtime a year. This availability level equates to 99.95% and is sufficient as a minimum. This is the standard based on the hardware and software components and a dedicated Ethernet Switch.

- **High Availability**—This server level is considered four 9s (99.99%) of availability, which equates to roughly 53 minutes of downtime a year on average. High availability is often desired in business areas with much less tolerance to downtime. This level is made possible by using redundant equipment, with redundant Ethernet switches, IP server interfaces (IPSI) in at least four out of five port networks (gateway connections), and duplicate control paths.

- **Critical Availability**—At this server level, five 9s (99.999%) is achieved. Critical server configuration is necessary in mission-critical business environments where downtime would be catastrophic. Critical level means less than five minutes of downtime per year. Typically the type of business that chooses this high standard would include hospitals and emergency response, contact centers, financial institutions, government agencies and other organizations that have lives, money, or national security on the line. Critical availability is achieved using the foundation defined for high availability while adding redundant expansion or ATM interfaces for bearer paths and duplicated center stage port network connectivity (PNC) or ATM switch PNC solutions as appropriate.

Note: The terminology "port network" is an Avaya term that defines network connections for bearer and control capabilities. Media Gateway is the terminology that is used in the VoIP environment to define similar characteristics.

Multi-Connect vs. IP-Connect

The S8700 can be set up as a multi-connect or IP-connect configuration. Multi-connect is used in environments where traditional port network connectivity is still being maintained. This means that the S8700s would be connected to other SCC1 or MCC1 equipment. This solution usually includes an IPSI board to make the connection to the EPNs. A center stage PNC solution or ATM PNC can also be used in larger environments to make this connection and scale the network.

IP-connect means that all bearer and control communications happen within the network architecture. For this to happen, the SCC1 and MCC1 are replaced with media gateways like the G600. An IPSI board is still used, but the PNC requirement is removed from the design. CLAN boards are used by the media server for call processing and control. The Media Processor (MedPro) board is used to perform the conversion from TDM to IP.

Interoperability

The S8700 and other media server and media gateway combinations can provide significant new functionality in the form of IP communication capabilities and converged network solutions. When connecting a new media server to an existing and traditional voice network environment, it is necessary to do a little homework. During the design process it is a good idea to review all hardware and software components for interoperability. It is possible to utilize many of the features and adjuncts available from the DEFINITY server (PBX). Some things to review include the vintage of the circuit packs for the IPSI, CLAN, and Medpro. It would also be worthwhile to verify the server software version. When connecting to adjuncts such as a Call Management System (CMS), check and verify the versions of software and hardware components for compatibility. It may help save time and money to provide the installation and support person the information they need to make things work properly.

S8500

A newer addition to the Avaya product line, the S8500 offers a midrange product for medium to large organizations. The S8500 is built on the same strong foundation of Linux and Intel processors, as well as the Communication Manager software. It provides connectivity for up to 2,400 analog, digital, and IP stations, along with 800 trunk links. It offers the same networking capability as the S8700, supporting 250 gateway connections for high scalability. It supports various gateway options, including the G350, G650, and G700, with additional configuration options of the G600, SCC1, and MCC1. The S8500 is configured as a simplex server, meaning that it is not designed with the dual (duplicate) server option of the S8700. However, it does offer survivability functions as well as design options for redundancy. The media server can provide 100,000 busy hour call completions.

S8300 Media Server with G700 Gateway

The S8300 offers many of the benefits of an integrated, converged network solution that works well in small to medium sized branch locations. It can serve from 40 to 450 analog, digital, and IP end stations and up to 450 trunks can also be connected. It provides full functionality of the Avaya Communication Manager Software, offering a robust level of features and service quality. The hardware includes a Layer 2 Ethernet switch that offers hot swappable, redundant architecture to prevent outages. The standard configuration is an S8300 server blade that is housed in the

G700 or G350 Media Gateway equipment. It also offers a stackable design for easy scalability, and can connect up to 50 media gateways (G350 or G700) for network expansion. The S8300 offers an impressive number of 10,000 busy hour call completions.

Survivability

When designing a remote/branch office location, it is necessary to understand the communication requirements in the unlikely event of an outage. For example, the branch office might require that call control and feature availability be maintained even if a connection is lost to the main server. Local Survivable Processor (LSP) is the solution that allows a local server to preserve call and feature capabilities until the connection to the main server is reestablished in a networked environment. This helps duplicate server functionality from the main office and prevents a single point of failure. This design scenario allows internal calls to be processed. To expand on that, local trunks may be connected in the event that a WAN connection to the main office drops. Local trunks allow call processing directly from the branch office. Other functionality may also be desired at the remote location in the event of an outage, such as messaging. The S8300 has an embedded Multimedia Messaging Application (INTUITY™ AUDIX), which can provide this functionality locally.

The S8300 Media Server with the G700 Media Gateway supports both standard and high-availability designs. Standard level fails within the 99.99 to 99.995 percent range of availability, and is considered the acceptable minimum for most business environments. For businesses that are more mission critical, the high-availability level is designed using duplicate network interfaces, as well as two or more IP connections to other servers (S8300 or other). This duplicate IP connectivity allows end devices to re-home in the event that server connectivity is lost, and maintain communication and feature support.

The benefit of using S8300 at branch office locations while maintaining one or more S8700s is that the media servers can be centrally administered from the main site. At the same time, new opportunities are available by extending IP functionality to branch locations. Selecting equipment designed with a local survivable processor means that the branch office can continue working even through a WAN failure.

S8100

For smaller office environments that need a strong and stable phone system, but are less concerned about availability issues or extensive scalability, the S8100 is a good solution. The S8100 is a server blade installed in a G600 Media Gateway. It has an Intel processor and Windows Server 2000 operating system. It connects the same number of analog, digital, and IP stations as the S8300, but only offers the option of connecting 300 trunks. It can complete 5,000 calls during a busy hour and can scale the network by connecting up to 10 additional media gateways (G600 or CMC1). The S8100 is an all-in-one system for small businesses by providing INTUITY™AUDIX embedded messaging, as well as Integrated Contact Center for call distribution services. Numerous other features and functions are also available for this powerful system to meet the demands of small organizations with the future in mind.

Media Gateways

A gateway is a device or service that provides protocol translation and access to another network. Gateways offer the ability to convert between TDM and IP packets. They also allow connectivity from a private network into the PSTN. Gateways are used to connect different stations (analog, digital, and IP), as well as the many different trunk types (analog, ISDN BRI/PRI, T1/E1, ATM, IP, etc.). Avaya uses the terminology of port network to define the services offered by a gateway. Therefore, a traditional PNC solution includes the functionality of gateway services. A gateway depends on a call server for processing. Its job is one of conversion or translation. Therefore, media servers and media gateways are combined together for complete functionality. The following are some of the Avaya gateway products, their characteristics, and the services they provide.

G350

The G350 is a scaled-down version of the G700 media gateway. It uses the same concept of media modules. It also supports the S8300 media server for internal call control functionality. Unlike the G700, the G350 is designed more as a standalone solution, offering the option of connecting up to 40 extensions. It has five media module expansion slots in addition to a one high-density media module slot. Analog, digital, and IP stations are supported, as well as analog, ISDN BRI/PRI, T1/E1, Gigabit Ethernet, and IP (H.323 or Session Initiation Protocol [SIP]) over frame relay or PPP.

G600

The G600 media gateway houses ten traditional TN-board expansion slots. It is most often combined with the S8100 media server board for internal call processing functions. Otherwise, it accesses another server for external call processing. It supports analog, digital, and IP phones like all the gateways. It also supports trunk links for analog, ATM, ISDN BRI/PRI, T1/E1, and H.323 or SIP over IP. It scales to four units within a port network.

G650

The G650 is a larger model based on the G600. It has 14 TN-board expansion slots. It has external call processing functionality from another media server. It offers the same station and trunk configurations. The major difference is that it scales to five units within a port network. The G650 is often connected to the S8500 or an S8700 media server.

IP Office

The IP Office is truly an all-in-one solution for data and voice convergence. It offers a router, Internet access, firewall, voice server, messaging, and conferencing in one piece of equipment. Designed for small and growing organizations that are planning for the future, the IP Office provides everything necessary to get voice and data up and running quickly.

As a voice server, the IP Office supports anywhere from 2 to 256 analog, digital, IP-based, or wireless stations. It also has a Contact Center component, which helps manage and distribute calls for up to 75 agents into voice, e-mail, or Web chat environments. It includes a messaging and auto-attendant component (Voicemail Lite or Voicemail Pro) that can be extended to a unified messaging platform and is integrated with the Microsoft Exchange e-mail application. Conference bridging allows businesses to keep in touch with up to 64 people per call and helps to reduce the costs associated with third-party call conference providers. Trunk options include up to 96 digital (T1, ISDN BRI/PRI) and 192 analog trunks.

For expansion purposes, an IP Office can be connected to a PBX or media server, making it an ideal candidate for a branch office. For organizations that do not have a centralized environment, up to 16 IP Offices can be connected to distribute converged networking functions. Regardless of the configuration, centralized administration is still possible from a designated location. Connectivity can be established through any WAN connection that supports IP communications, such as frame relay, PPP, and Gigabit Ethernet. When using frame relay, VoIP services should be provisioned to help ensure a level of quality for voice communications.

On the data side, there are 10/100 Mb Ethernet LAN ports that can be used to connect PCs, servers, phones, printers, and other end devices. WAN ports provide access to the Internet, while the internal firewall helps secure the network. The IP office can also be used for remote access using dial-up or VPN connections via the Internet.

Supported Telephones

The Avaya Communication Manager environment supports analog, digital, and IP telephones. Use of analog and digital phones allows customers to maximize their investment, have backward compatibility, and still progress into the future. IP phones offer new hardware- and software-based options using H.323 or SIP. IP capabilities also offer more options for types of end devices, from traditional desk phones, to software on a PC, or a hand-held device.

Analog

The 6200 series analog phones offer basic business class features. They provide the flexibility of connecting a laptop, PC modem, or fax through a data-jack (RJ11) on the side of the phone. They also offer redial, message waiting, flash, positive disconnect, and hold functions. Different sets provide additional features, such as personalized ring, programming keys, hands-free speaker phone, and mute.

Digital

The Avaya Digital Communication Protocol (DCP) phones are designed for various business uses, but offer such a diverse group of features that they are often used in contact center environments as well. They connect to Avaya Communication Manager servers, to take full advantage of the diverse feature set. The 2420 series telephones are two-wire digital sets that are ergonomically designed and offer an LCD. They offer 24 feature/call appearance buttons that can be programmed as appropriate for individual users. In addition, the display language can be set at the station, regardless of the switch language selection. Other key features include speed dial directory, extension download, and numerous other options.

The 6400 series is another line of DCP sets available from Avaya. This series is also fully integrated with Avaya Communication Manager feature sets. In addition, the series offers group listening, which allows a person to use the handset, as well as the speaker, for group conversations, whisper paging to let someone know a call is waiting, group paging, call timer, one-touch operation, and many other options.

The server instantly recognizes the DCP digital phones, which are available to use as soon as they are plugged in. These phones can be upgraded to protect the investment and operability into the future. They also demand less processor time per call from the Avaya Communication Manager server as compared to the IP telephones.

IP

The Avaya Communication Manager server environment supports a number of IP telephones. They vary from a standard business phone up to a Web-enabled browser display model. IP phones are offered in two different types: hardphone and softphone. The hardphone is a desktop-based set with similar look and feel of a digital station. The softphone is software that runs on a PC desktop or PocketPC. Regardless of the format, an IP phone uses the IP protocol suite and can work in a data network, as well as the Internet. IP phones must have an IP address, subnet mask, and other parameters assigned before they can function and be identified on the network. They offer extensive versatility of applications and integration capabilities with other IP applications. IP telephony requires additional components, such as encryption for security, network power and protection from viruses, and denial of service attacks. Other important considerations are quality of service and the impact that network issues, such as jitter and delay, have on voice traffic. Avaya has taken all this into consideration and designed phones that will provide security, protection, power over Ethernet capabilities, industry standard IP applications and communication methods H.323 and Session Initiation Protocol (SIP), and administrable quality of service parameters (Diffserv code points, 802.1p and 802.1Q).

IP phones take some time to become active. When first plugged in, the IP phone must locate a Dynamic Host Control Protocol (DHCP) server and request an IP address. Then if configured, it will look for a Trivial File Transfer Protocol (TFTP) server for a firmware upgrade, or possibly run a script for additional configuration parameters. After these processes have been performed, the phone is available for use. IP phones also demand more processor resources from the Avaya Communication Manager server because of the additional services required to make these phones operational on the network.

Wireless

There are a number of different wireless telephone options that offer mobility solutions for people who are regularly away from the office, but still need the features and functionality of a business class phone system. Wireless telephones are offered as digital devices. These use the 900-Mhz unlicensed frequency to create a micro-cell network within a building. Another wireless phone option includes two-way radio functionality as well, which is handy for timely conversations.

IP wireless products are also available, including an IP softphone application that runs on standard PocketPC equipment, and a full-feature wireless IP phone that comes in either a lightweight professional style or a rugged model for more institutional and industrial environments. The IP wireless phones support the IEEE 802.11a, b, and g wireless LAN standards using unlicensed frequencies. These different wireless phone types work with the Avaya Communication Manager architecture allowing service by either a media server or DEFINITY.

Activities

1. Which availability method achieves five 9s (99.999%) reliability for the S8700?

   a. High

   b. Standard

   c. Extended

   d. Critical

2. Which two components are often combined together for a complete IP solution of media gateway and media server? (Choose two.)

   a. S8300

   b. S8700

   c. G700

   d. IP Office

3. What component is necessary when a branch office decides that it requires the media server to continue processing calls, even if a connection to the main office is down?

   a. Redundant link

   b. Local Survival Processor

   c. Live processor

   d. Hot swappable processor

4. Which two components can a media gateway translate between? (Choose two.)

   a. TDM

   b. ATM

   c. IP

   d. DHCP

5. What software component within Avaya's product line is responsible for providing all call processing, communication, and feature functionality?

   a. Avaya Communication Manager

   b. Avaya IP Communicator

   c. Avaya ECLIPS

   d. Avaya DEFINITY Enterprise Communications

Extended Activities

1. You have been tasked with designing a converged network for a small company that is expanding rapidly. The company has 25 employees now and wants a solution that offers them data and voice capabilities, as well as voice mail servers. What product solution do you propose?

   a. A DEFINITY with an Ethernet switch

   b. An IP office

   c. A G650 media gateway

   d. A DCHP server with firewall software

2. A company is expanding into a new location. They want to use their existing phone equipment at the main office since it is a new S8700 media server. They want to make sure that they will be able to function if the WAN link goes down between the two offices. What components do you suggest to provide a solution? (Choose two.)

   a. An S8300 media server with G700 media gateway, and LSP

   b. A S8700 with critical availability

   c. Local trunk links

   d. Two S8300 media servers, one to backup the other

3. A customer is choosing between digital (DCP) telephones and IP telephones to connect to an S8700. Which of the following are characteristics of the IP phones? (Choose two.)

   a. The IP phones take a little longer to be recognized and become available than digital phones.

   b. The IP phones demand more processing capabilities of the server.

   c. The IP phones are more user friendly than the digital phones.

   d. The IP phones require less processing capabilities of the server.

4. A branch office has a G700 media gateway. What offers the local call processing capabilities for this hardware component?

   a. An S8300 server blade installed in the G700

   b. A G350 server blade installed in the G700

   c. A media server processor installed in the G700

   d. An expansion port network processor installed in the G700

5. A customer has requested that their offices be networked together for voice and data communications. They have a frame relay link and IP Office hardware installed in both locations. Will it be possible to network these two offices together?

   a. Yes, because frame relay transports IP information for both voice and data.

   b. No, because frame relay does not offer a reliable transport method.

   c. No, because frame relay is not supported for networking IP Office equipment.

   d. No, because two IP offices cannot be connected together across a WAN link.

Lesson 6—PBX Networking

In the realm of telephony, a private communications network is an interconnected group of communications systems (PBXs). People within each system, called local users, can exchange voice and data with other individuals at communications systems in the network, called nonlocal users. The systems in a private network may be located on the same campus, or separated by thousands of miles. This lesson reviews general concepts necessary to understand how to network PBX systems together.

Objectives

At the end of this lesson you will be able to:

- Explain what a private communications network is, and how it benefits an organization

- Explain how tandem switching works

- Describe and compare three common configurations of private communications networks

 **Key Point**

A private PBX network can carry long distance traffic over private lines.

Private Communications Networks

When organizations have several locations in different regions, they often implement a PBX system at each branch office, then use the PSTN to handle communications between those systems. This approach is shown on the Public Communications Network Diagram.

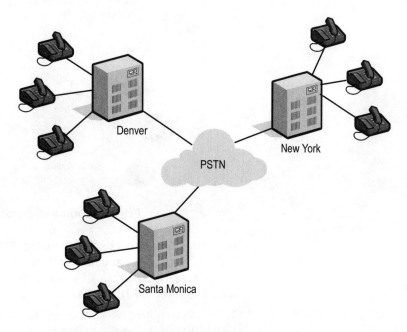

Public Communications Network

This arrangement works well, as long as the volume of interlocation telephone traffic is fairly low. For companies that require a higher level of long distance traffic, it makes more sense to connect regional offices with private leased lines. This approach is illustrated on the Private Communications Network Diagram.

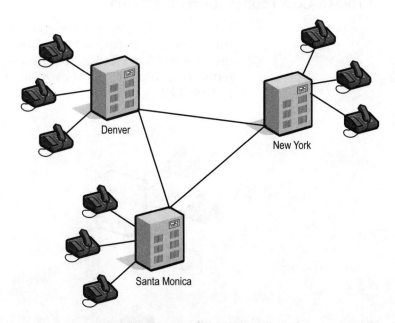

Private Communications Network

In a private network, the separate communications systems are linked by private transmission facilities. These lines/trunks may be analog tie trunks, T1-emulated tie trunks, or ISDN-Primary Rate Interface (PRI) trunks.

When a company needs the control of a private network, with the flexibility of a public network, the answer may be a software defined network (SDN). The AT&T SDN is a flagship product in the AT&T nodal architecture. SDN gives customers a corporate virtual private network (VPN) implemented over the facilities of the AT&T public switched network.

SDN is a customer's VPN that resides in the AT&T 4ESS™based switched Worldwide Intelligent Network. It provides features and management capabilities that are usually not found in private networks, such as customized routing, advance numbering plans, call screening, authorization codes, remote access, security codes, and customized billing. The service fully supports analog data transmission at up to 28.8 Kbps and end-to-end digital data transmission at 56/64 Kbps.

SDN is compatible with most private networks and PBXs, thus protecting these existing investments. Since SDN does not require a sophisticated PBX base, businesses can choose dedicated access (T-carrier or ISDN) or dial-up access. Traveling users can get dial-up access to the SDN by providing a valid authorization code.

No matter how it is implemented, private networks are distinct from the worldwide PSTN. A private network carries calls within an organization, while PSTN lines and trunks carry calls to local and long distance parties outside the organization. When properly implemented, a private PBX network can achieve significant cost savings and improved efficiency, increased user satisfaction, and improved security.

Cost Savings

Private PBX networks can reduce both toll charges and the costs of leased lines.

- All calls between company locations bypass the PSTN, eliminating toll charges. For example, if a company has locations in Dallas and Cincinnati, all calls between those cities are carried by the private network. Users can dial any company extension, anywhere in the network, just as they would dial an extension on their own local system. Using Automatic Route Selection (ARS), the system transparently routes the call over the correct trunks or service providers.

- Calls to destinations outside the company can use the private network to reduce toll charges. For example, consider a company with locations in Seattle and Orlando. A call from Seattle to Miami first travels free, over the company network, to Orlando. Then the Orlando PBX transfers the call to the PSTN for routing to Miami. The company is only billed for long distance service between Orlando and Miami.

- Leased trunks are cheaper than full-service switching. A company can order a point-to-point T1 circuit from a service provider, then use system programming to set it up for tandem ISDN-PRI services. The telecommunications service provider provides amplification (repeaters) for PRI trunks when the distance between networked systems is great enough to distort signals. However, the service provider does not supply higher-cost switching services.

Improved Efficiency

Private PBX networks can also improve efficiency.

- Leased trunks can be used for much more than voice calling. A company can tailor its use of PRI B channels by using drop-and-insert equipment that allows fractional use of T1 channels for data/video communications between sites, while keeping the remaining T1 channels for PRI voice or data traffic. T1 channels can support a mix of T1-emulated tandem tie trunks for voice or data communications at 56 kilobits per second (Kbps) per B channel, while allowing data transfers over two or more channel-bonded B channels.

- When appropriate, a company's incoming call traffic can be spread over the entire private network. By configuring calling groups, overflow calls from one location can be routed to other locations, increasing the number of coverage points and sharing personnel and resources between systems.

- A centralized voice messaging system can provide additional savings by not requiring a separate voice messaging system at each location in the private network.

Enhanced User Satisfaction

When a private PBX network is implemented to function as a single system, it offers its users simpler operation and a wider range of features.

- A Uniform (or Universal) Dial Plan establishes a single numbering system for all company users. Within the private network, all calls appear to be local.

- When switches are linked with ISDN connections and the Q.SIG inter-switch protocol, the network can offer the full range of ISDN and Q.SIG supplementary services, such as calling party ID or call forwarding.

DEFINITY Server, MultiVantage software, supports UDPs of up to seven digits in length for local extensions, including stations, data modules, call vectoring, and agent login IDs. PBX administrators have the flexibility to administer dial plans between three and seven digits in length, and the MultiVantage software platform supports mixed digit lengths in the same dial plan. Q.SIG is the required networking protocol if the dial plan between networked switches exceeds five digits in length. DCS only supports three- to five-digit dial plans, whereas Q.SIG supports three- to seven-digit dial plans.

Tandem Trunks and Tandem Switching

The leased lines that connect private PBX systems can be called private network trunks, because they enable private networks. They are also called tandem trunks, because a private communications network allows tandem switching.

Tandem trunks can be point-to-point services such as T-carriers, switched services such as ISDN-PRI, or virtual circuits. T-carriers or ISDN-PRI are popular choices for most corporate networks, but the final choice depends on the services your network needs. If your only requirement is multi-channel connectivity, then a 24-channel T1 line is usually more economical than 23-channel ISDN-PRI. For example, simple voice and data connectivity often requires only T1 channels. But if your business needs the rich feature set of ISDN, then the extra cost of that service is justified.

In the PSTN, a tandem switch is an intermediate switching point. It is a CO switch that does not serve a call's destination, but relays the call to the switch that does serve the destination.

In the same way, PBX systems can serve as intermediate switches in a private network. When one PBX receives a call, then forwards it to another PBX, it is serving as a tandem switch.

A tandem-switched call does not necessarily terminate at another system directly connected to your own. It may travel over the private network via an intermediate networked system. Furthermore, a nonlocal system may direct the call to a PSTN facility and then to someone located outside the private network.

Network Configurations

This section looks at some of the ways private electronic tandem networks can be configured, and discusses the differences among them. It does not illustrate all possible ways private networks can be connected.

Series

In the Series Configuration Diagram, systems are arranged in a line with no central system acting as a hub. In this arrangement, PBXs A and D are called peripheral systems, because neither connects to more than one switch in the private network.

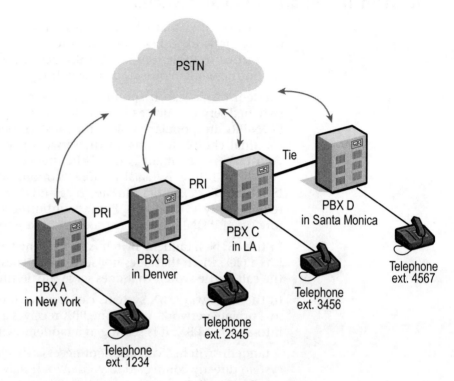

Seires Configuration

This configuration requires that the private network sites be connected using either a delay-start tie line (analog or T1-emulated) or a digital T1 circuit that has been programmed for PRI.

To make a call from PBX B in Denver to PBX D in Santa Monica, a user at extension 2345 dials 4567. The call travels over tandem trunks, through PBX C, without using the PSTN to provide switching services. In this configuration, PBXs B and C can serve as tandem switches.

Using another aspect of tandem switching, the LA user at extension 3456 can employ ARS to dial a New York number that is outside the private network. For example, when the user dials "9-1-212-555-2468," the initial 9 is the ARS code that tells PBX C to route the call to PBX A over tandem trunks. At PBX A, the call goes out over a facility connected to the PSTN as a local call from PBX A.

These two calls have the following features in common:

- All or part of each call is carried over tandem trunks.

- The calls are routed seamlessly from a source extension to a destination extension.

- The cost of tandem-switched calls is substantially lower than the cost to make the same calls using telecommunications service providers and the PSTN.

- The users dial the calls normally. To make "inside" calls, the user simply dials an extension number. To make long distance ARS calls, the caller uses a System Access (SA) button. ARS and Universal Dial Plan (UDP) routing are programmed to take advantage of all PSTN facilities in the network.

Star

The Star Configuration Diagram illustrates the second possible arrangement for a private network.

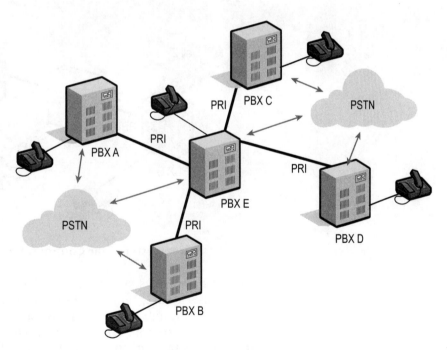

Star Configuration

In a star configuration, all network-routed calls pass through a central hub system. The Avaya DEFINITY Server and DEFINITY ProLogix Solutions systems are examples of devices used as hubs. The hub may normally terminate individual stations as well as external switching facilities. Star system users make calls in the same way as they would in a series configuration.

Three-System Star

The Three-System Star Configuration Diagram illustrates a simpler star configuration that consists of only three PBX systems. It may look like a series, but all network-routed calls pass through PBX B, just as they do in the larger star configuration. An example of a PBX used in this configuration is a Merlin Magix PBX.

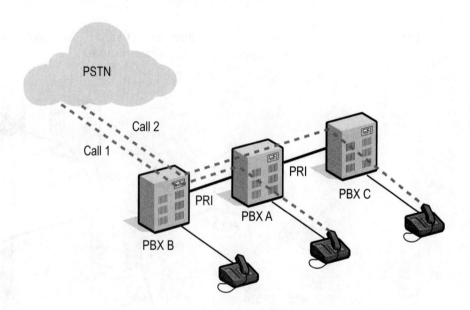

Three-System Star Configuration

This diagram presents specific PSTN facilities as well as tandem trunks. Calls 1 and 2, represented by dotted lines, enter the private network through PBX B. Call 1 is routed directly to an extension on PBX A. To route Call 2 to an extension on PBX C, PBX A must serve as a tandem switch. In this way, the three-system star combines features of series and star configurations.

Both DID and ISDN-PRI outside facilities permit this type of routing. However, the types of tandem trunks that connect the systems, as well as the lines/trunks connecting to the PSTN, also affect the decisions made about private network configurations.

PBX Security

Toll fraud is a constant threat to PBX systems. Remote access, or DISA, can provide a "back door" for hackers to freely access outbound long-distance lines. Auto attendants may make available menu choices that give hackers access to private PBX services. Modems on service ports provide yet another avenue for hackers to steal services. Because of these vulnerabilities, the system administrator must implement sound security practices to prevent theft of valuable company resources.

To increase voice system security, you can take the following precautions:

- Disable unused login accounts.

- Restrict remote access to only those users who absolutely need it.

- Assign barrier and authorization codes to remote users. Barrier codes allow access to a feature, while authorization codes allow or prevent outbound calls.

- Use Classes of Restriction (CoR) and Facility Restriction Levels (FRL) to control access to PBX services and trunks.

- Use Time-of-Day routing to prevent after hours outbound calling.

- Disable trunk-to-trunk transfers.

- Disable remote access ports or remote access services.

- Monitor system logs, including CoR and FRL usage.

IP PBX Security

IP PBXs open the voice network to a whole new realm of security threats. Viruses, denial of service attacks, spoofing, worms, and all other sorts of attacks are possible when a PBX is connected to the data network.

In addition to the traditional toll fraud prevention methods mentioned above, the following techniques used on data networks can be applied to IP PBX systems:

- Place voice systems on VLANs separate from the rest of the data network.

- Place voice systems on separate subnets.

- Use firewalls to allow only authorized traffic access to the PBX.

- Use encryption on VoIP signaling and media streams.

- Use Layer 2 and 3 security protocols to secure client connections to the voice server (IPSec, 802.1x, RADIUS, and so forth).

- Regularly apply patches and security updates.

- Monitor server logs for suspicious activity.

Avaya markets a suite of products designed to secure access to IP data and communications systems. Called the Avaya Access Security Gateway (ASG), this product line includes the following components:

- **ASG Guard**—Multiport access control units connected between the systems and adjuncts and the rest of the network.

- **ASG Key**—A small key device that generates a one-time use system access code.

- **ASG Guard II**—A built-in VPN and firewall technologies.

- These products are designed to augment the system or adjunct's built-in security capabilities.

Activities

1. What is the basic difference between private and public networks?

   _____

   _____

   _____

   _____

2. What is the difference between a private line and the world-wide PSTN?

   _____

   _____

   _____

   _____

3. A company has four locations throughout the United States. Each location has a PBX and connects to the other locations by means of the PSTN. When would it make sense to use leased lines for the PBX-to-PBX connectivity instead of the PSTN?

   _____

   _____

   _____

   _____

4. What are the tradeoffs between using the PSTN and leased lines?

   _____

   _____

   _____

   _____

5. What is DISA?

    a. A PBX feature that allows an outside caller to dial directly into the PBX system and use the system remotely

    b. A process by which a PBX routes calls directly to a particular extension

    c. A PBX function, that answers incoming calls, plays a greeting and instructions, then allows callers to directly dial an extension

    d. None of these choices describe DISA

6. Name three ways you can safeguard a PBX. (Choose three.)

    a. Disable trunk-to-trunk transfers

    b. Disable unused login accounts

    c. Monitor system logs, including CoS and FRL usage

    d. Disable all incoming calls

7. At what OSI model layer(s) are security protocols run to protect client connects to the voice server?

    a. Layers 1, 2, and 3

    b. Layer 3

    c. Layers 2, 3, and 4

    d. Layers 2 and 3

Extended Activity

A company has six locations within a single metropolitan area. Each location is large enough to have its own PBX system. Draw each of the configurations discussed in this lesson (series and star). Discuss how you would determine which configuration would be best for this scenario.

Lesson 7—Basics of Automatic Call Distributor

An ACD is optional switch (PBX) software that automatically distributes high-volume, incoming call traffic to available extensions within groups of call center agents. An ACD provides many flexible ways to distribute incoming calls.

An ACD works by automatically connecting incoming calls to specific hunt groups, called "splits" or "skills." Based on customer requirements, incoming calls are distributed to available agents within the most appropriate split/skill.

Objectives

At the end of this lesson you will be able to:

- Briefly explain how an ACD works

- Describe the different types of split queues

- Name and describe the methods of distributing calls to agents

- Explain why it is important to detect and minimize abandoned calls

 **Key Point**

An ACD can be configured to provide very flexible call routing.

Agents and Splits

Any entity that answers calls, a live person or automated voice response port, is called an agent. Agents are often organized into groups, called splits. Each split sells similar products or provides the same types of service.

Each ACD split can be assigned a split supervisor. The supervisor uses split/skill administration features, either as a component of the PBX software or as a part of the ACD CMS, to assign agents to splits, move agents from one split to another in response to changing call volumes, monitor agent performance, and provide assistance if necessary.

One of the split supervisor's most important jobs is to determine the optimum staffing level for each time of day, week, or year. To do this, the supervisor must balance the call center's need for cost reduction against the customer's need for speedy service.

It is simple to lower costs by reducing the number of trunks from the CO, and hiring fewer agents to answer calls. However, the call center loses revenue if customers abandon calls because of the difficulty of reaching an agent. Therefore, call center management must determine the number of trunks and agents that will minimize costs and maximize customer ability to purchase goods or services.

Some ACDs are equipped with software that uses call traffic statistics to help forecast the number of calls agents can expect at each half hour of the day. By using this information, management can refine its staffing plans, and adjust ACD settings accordingly.

What an ACD Does

As illustrated on the ACD Configuration Diagram, incoming calls arrive by means of a number of different trunk types and services. The ACD answers calls, then places them into a queue. A queue is a holding area for calls waiting to be answered, usually in the order in which they were received.

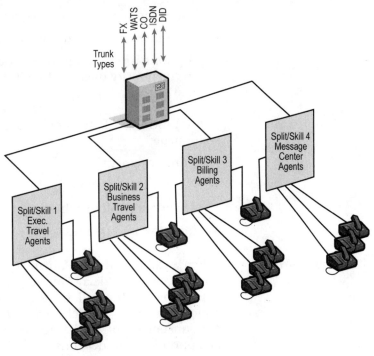

ACD Configuration

Depending on the business of the call center, the ACD may use one queue or several different queues (one for each split). The business shown in the diagram has implemented four splits, each with its own queue. The ACD identifies the destination split of each call, based on Automatic Number Identification (ANI) or Dialed Number Identification Service (DNIS)—used to route the caller based on caller location—or a caller's selection from an IVR menu. Each split can also receive incoming calls through DID processing, automatic-in processing, or both.

Automatic-In Processing

Through switch administration, each automatic-in trunk group can be assigned to an ACD split. Then all calls that come in on an automatic-in trunk group are automatically routed to the assigned split queue.

DID Processing

Some ACD systems enable customers to dial directly to various splits or individual agents. For DID processing, each DID extension is assigned to a split. All calls for one of these DID extensions are routed to the appropriate split queue.

Split Queue Processing

Each split can have one or two queues:

- A normal queue is required for any split.

- A priority queue is optional. The ACD first distributes all calls in the priority queue before it distributes any calls in the normal queue.

The following ACD features and functions are available to any split queue:

- Announcements

- Answer supervision

- Abandoned call search

- Intraflow and interflow

Announcements

When a call is put into a queue, the caller may hear one or more announcements (delay messages, music, or advertisements), depending on the call treatment assigned for the split. These announcements are delivered by either external or internal announcement units. The number of calls that can be queued to an announcement depends on the size of the PBX switch.

Internal announcements are delivered by a multichannel integrated announcement board in the PBX; a call receives an announcement only when it connects to one of the announcement channels. Therefore, all calls wait in a single queue to access a channel on the announcement board, regardless of the split announcement they are waiting to receive. The same announcement can be delivered over multiple channels. Announcements are delivered on demand, so each call that connects to a channel receives an announcement immediately, and does not have to wait for the announcement to finish and start again.

When caller traffic increases, the system may have problems delivering announcements. Callers may hear delayed announcements or none at all. At this point system upgrades due to increased caller traffic may be in order. To supply announcements, the PBX requires tone detectors and announcement boards of sufficient number and capacity to handle busy hour traffic at the desired grade of service (GoS).

Tone detectors, integrated with call classifiers, collect the digits required by the vector treatment before the call is transferred to the appropriate announcer queue. Each active call ties up a tone detector/call classifier board resource (port) for the amount of time required to collect the digits from the caller. If a call classification resource is unavailable to collect digits from the caller, the caller may experience periods of silence waiting for a detector to become available, or the call may skip the digit collection step altogether. Likewise, if an inadequate number of announcer boards (either TN750C 16-channel Recorded Announcement boards or 30-port TN2501AP Voice Announcement over LAN [VAL] boards) are available, once the call is routed to the announcer queue, the caller may not hear the announcement or it will be delayed. Analysis of the system will indicate whether more call classification and/or announcement resources are required.

Answer Supervision

Answer supervision is a signal sent by the PBX switch to the serving CO. This signal tells the CO that an incoming call has been answered, so the CO can begin tracking toll charges for the call (if they apply). Answer supervision is sent just before a call connects to an agent's voice terminal, music, or an announcement.

Abandoned Call Search

A call is abandoned if the caller hangs up before the call is connected to an agent. A call can be abandoned at any time while in a queue, up to the moment an agent picks up the telephone.

Abandoned calls represent lost sales or damaged good will. Thus, splits should be staffed so that calls do not have to wait in a queue for an unreasonable amount of time, and announcements can be used to persuade the caller to wait until someone answers the call.

When a caller abandons a call just after answer supervision has been sent, a "ghost call" can be connected to an agent. This problem can occur because, after a caller hangs up, some COs wait 2 to 25 seconds before sending a disconnect signal to the PBX switch. Ghost calls are a problem because they waste agents' time, and can delay or prevent other calls from connecting to an agent.

To minimize this problem, abandoned call search can be assigned to specific trunk groups on some switches. With abandoned call search, the switch checks the incoming trunk before delivering an ACD call to an agent. If the trunk is on-hook at the CO (call has been abandoned), the switch releases the trunk and does not deliver the call. The switch only delivers a call to an agent if the call is still in progress on the trunk.

Intraflow and Interflow

Intraflow and interflow allow calls to be redirected to another split, a different local destination, or a remote destination. Redirecting calls to a local destination is called intraflow. Redirecting calls to a destination outside the local switch is called interflow.

Call Distribution Methods

Once calls are in a queue, the ACD's distribution methodology determines how it distributes calls to agents. An ACD can distribute calls to agents in a split in one of four ways:

- **Uniform call distribution (UCD)**—The ACD assigns each call to the agent that has been idle for the longest time. This approach distributes the workload evenly across the split.

- **Direct hunting, also called "linear" or "top-down"**—The ACD distributes calls by working through a list of agents, always starting at the top. This means agents at the top of the list are always busier than those on the bottom.

- **Circular hunting**—Similar to top-down hunting, the ACD assigns calls to each agent in turn. However, an ACD that uses circular hunting keeps track of its current position in the list. It behaves like a card-game dealer who gives each player a card in turn.

- **Expert agent distribution (EAD), or skills-based routing**—The ACD assigns each call to an agent who has one or more of the skills necessary to handle the call.

Split supervisors can also set the ACD to overflow calls to another group of agents, should the call volume be more than one group can handle within a predetermined time. Real-time and historical reporting features provide supervisors with the information they need to quickly adjust the distribution of calls to improve service.

Vectors

Vectors are call routing scripts used by the PBX or ACD to determine how the ACD treats a call. For example, a bank can provide a single toll-free number for customers to call concerning the status of their mortgage applications. A vector directory number (VDN), assigned to an incoming trunk group or passed to the switch via DID or DNIS, allows the ACD to apply a call treatment based on the called number, the calling number, time of day, or some other criteria.

When the inbound trunk accesses the VDN, the VDN calls a vector script, which tells the ACD what to do with the call. In our banking example, this script could call up an announcement prompting the caller to select the service required from a menu of spoken choices. The caller's menu choice could call up another vector, or route the call to a specific split. Multiple VDNs may call up the same vector, enabling the same series of call treatment steps to service calls from different numbers or locations.

After the call reaches an agent, the agent can see information about the call, such as the name assigned to the VDN. This enables the agent to quickly determine the service or response the call requires. Returning to our banking example, suppose the caller chose the menu item requesting loan status. The vector may be named "Status Inquiry" so that when the agent sees this name on the telephone display, the agent knows the customer wants loan status information. The agent can then access the correct database screen to quickly locate the caller's information.

Vector Flow

The ACD runs the vector script sequentially, from the first step to the last. The process ends automatically at the end of the script. Vector processing terminates when a call has left the vector, such as when it is passed to an agent's workstation. Several call flow methods are used:

- **Multiple split queuing**—Queues a call for up to three splits
- Intraflow
- Interflow

- **Look-Ahead Interflow (LAI)**—Reroutes calls to the local or remote location that can best handle the call. Used with ISDN-PRI service.

- **Best Service Routing (BSR)**—Available on a single or multi-site network, BSR allows the PBX to analyze the call and identify the split or skill best able to service it.

- **Adjunct Routing**—Uses the ASAI link (discussed in Module 2) to pass caller information from the PBX to an adjunct ACD. Based on the ACD's database, it routes the call to the best split or skill.

Vector Management

Although writing vectors is easy after some practice, problems do arise. Although we do not have space here to cover all the nuances involved in writing vectors, the next section briefly discusses some techniques that may help you trace vector problems.

Call Flow Problems

Avaya DEFINITY vectoring allows for up to 1,000 steps (3,000 with advanced LAI), so errors in the script can occur. Several problems can arise as a result of an error in the call flow:

- **Vector gets "stuck"**—This occurs if no default treatment is included in the script.

- **Skipped steps**—This is caused by application timeouts, invalid or unavailable resources, or ASAI or PN link failures.

- **Unheard announcement**—This is caused by a failed, missing, or misconfigured announcement board.

- **Delayed announcement**—This is caused by a full announcement queue or all ports or resources are busy.

- **Incorrect agent display**—This occurs if a vector passes incorrect call information to the agent locally or over an adjunct link.

Several methods are available for diagnosing vector problems. First, if a vector causes an error, the PBX records it as a vector event. Use the `display events` command to view the Event Report.

The `list` commands show historical information pertaining to ACD operation. For example, the command `list vector` shows at the supervisor terminal a list of existing vectors. The `list VDN` command shows all the VDNs associated with a vector. (Remember that multiple VDNs can point to the same vector.) The `list usage vector` command shows where in the PBX a particular vector is used. This could help trace which VDN initiated the call to an agent. The **list trace vdn** and **list trace vector** commands show how a particular VDN or vector behaves. Issue these commands to observe how the system processes a call, thus locating one or more faulty steps in the vector.

Another method for tracing call problems is to manually execute portions of the script to locate the step or steps where the problem occurs. By placing calls to portions of the flow, you can narrow down the cause of the vector problem to a related series of steps.

Activities

1. A call distribution method that allows the ACD software to check the split's/skill's agent numbers in the administered sequence, starting with the number of the last agent connected to a call is called _____.

2. One call distribution method allows ACD software to check the split's/skill's agent numbers in the administered sequence until it finds an agent with an available extension. It then routes the call to that extension. This method is referred to as

   _____.

3. One call distribution method allows an ACD to find the agent extension that has been idle for the longest period of time and route the call to that agent's extension. This method is referred to as _____.

4. A call distribution method based on agent skill is called

   _____.

5. An adjunct that collects call data from a switch resident ACD, provides call management performance recording and reporting, and can also be used to perform some ACD administration is called _____.

6. A group of trunks/agents selected to work together to provide specific routing of special purpose calls are referred to as

   _____.

7. A call collection point where calls are held until a split/skill agent or attendant can answer them is called

   _____.

8. A group of extensions/agents that can receive standard calls and/or special purpose calls from one or more trunk groups are referred to as _____.

9. What does a tone detector/call classifier do?

   a. It makes sure the line has a dial tone and can call the dialed number.

   b. It collects the digits punched in by a customer and routes the call.

   c. It detects line dial tone and routes the call according to the phone number dialed.

   d. It routes calls based on whether the number is local or long distance.

10. Name three vector call flow methods. (Choose three.)

    a. Multiple split queuing

    b. Vector management

    c. Look-Ahead Interflow

    d. Adjunct routing

Extended Activities

1. Based on the following types of organizations, what do you think would be the type of service each split group would perform? Also, indicate whether you think the agent would be a human or an automated voice response.

    a. Customer support from a bank for providing customer account information

    b. Customer support from a bank to discuss business and home loans

    c. Customer support from a manufacturer of computer networking and telephony hardware and software products

    d. Technical support for a provider of Internet service

2. Discuss how much time you would expect to wait for help for the following scenarios:

    a. To receive technical information for a software operating system vendor

    b. To receive help installing a new appliance for your home

    c. To receive information on your bank balance

    d. When calling 911 or another emergency service

Summary

In this module, we saw that a typical business PBX performs most of the same functions as a telephone company CO switch. At it simplest, it accepts calls coming in to one side, and connects them to available lines on the other. However, most PBXs support a long list of additional features, such as conference calling, call waiting, and call hold (even if most users never take advantage of them).

One of the most important features of a PBX is an ACD. Many businesses, especially call centers, could not exist without extremely sophisticated ACDs. These programmable systems automatically route calls to individual agents, or groups (splits) of agents, based on the sources of calls or facts about callers.

One of the most sophisticated digital PBXs is the Avaya DEFINITY Server and Media Server running Avaya Communications Manager (ACM). ACM's modular design allows for a wide range of flexible configurations. It is particularly good for implementing a wide area telephone network for a single, multi-location organization.

The DEFINITY Server's physical installation requirements are fairly typical of all complex electronic and computing devices. In general, systems such as DEFINITY Server require cool, dry air, consistent electrical power, and protection from electrical or radio interference. The physical requirements of telephone system technicians, bright light and generous working space, are just as important.

Because a PBX is essentially a small CO switch, several PBXs can be connected to form private communication networks. When linked with dedicated lines such as T1 or over an IP network, each PBX can serve as an intermediate switching point, relaying calls between switches that are not directly connected. This function, called tandem switching, allows large organizations to switch calls across a continent, totally bypassing the long distance telephone switching system.

Avaya offers a number of products in the Enterprise Class Internet Protocol Suite (ECLIPS) line. Avaya Communication Manager is the software that provides functionality for the entire line of communication equipment. The S8700, S8500, S8300, and S8100 media servers provide availability, reliability, and call processing capabilities fit for an organization of any size. The G350, G600, G650, and G700 Media Gateways allow users to continue to use their legacy DCP and analog telephones in concert with new,

IP-based hardphones and softphones. IP Office is an all-in-one solution providing sophisticated voice and data networking capabilities for small to medium businesses. ACM servers and media gateways support a number of telephones, from traditional analog and digital desk sets, to high-tech IP and wireless options.

Module 2 Quiz

1. Which of the following industries would most likely use a call accounting system for detail reporting for billing purposes? (Choose two.)

   a. Law firm

   b. Manufacturing firm

   c. Accounting firm

   d. Retail department store

2. Another name for a call accounting system is:

   a. SMDS

   b. SMDR

   c. Authorization codes

   d. ACD

3. What are the uses of the TDM bus slots found in a digital communications switch? (Choose two.)

   a. Carry switched voice, data, and control signals among port circuits

   b. Carry switched voice, data, and control signals among port circuits and the SPE

   c. ACD and PBX

   d. DS1 and ACD connectivity

4. Information enters and leaves a digital switch such as the DEFINITY Server through which of the following?

   a. Data ports

   b. Expansion ports

   c. Port circuits

   d. PNs

5. An accounting firm would like to track customers by the amount of time spent working with them on the telephone. Which items will provide the organization the needed information? (Choose two.)

    a. Authorization codes

    b. Call accounting

    c. Account codes

    d. CoS

6. The PBX feature that allows a user to resume a telephone call from another extension is:

    a. Call pickup

    b. Call park

    c. Call waiting

    d. DAU

7. A PBX becomes necessary when a company grows to:

    a. 5 to 10 lines

    b. 8 to 12 lines

    c. 12 to 18 lines

    d. 20 to 50 lines

8. Which of the following two configurations are used for PBX networking?

    a. Star and mesh

    b. Star and series

    c. Star and bus

    d. Ring and star

9. The PBX feature that allows a user to answer another telephone by means of his or her telephone is:

    a. Call pickup

    b. Call park

    c. Call waiting

    d. DAU

10. The PBX feature that allows a caller to hear messages while on hold is:

    a. Call pickup

    b. Call park

    c. Call waiting

    d. DAU

11. A customer wants calls answered by an office manager during normal business hours, but after hours calls are to be redirected to a different extension. Which feature should be used?

    a. CoS

    b. Call forwarding

    c. Line assignment

    d. Call pick-up

12. Which of the following is a description of PBX traffic flow?

    a. Number of CO lines into a PBX

    b. Number of incoming and outgoing calls

    c. Number of e-mail addresses

    d. Number of voice mail boxes available

13. A customer wants to know which employees are making the least number of calls. Which feature is needed?

    a. Call center

    b. UDP

    c. CDR

    d. ARS

14. The PBX feature that allows callers to use the telephone touchpad to access information is known as:

    a. IVR

    b. DID

    c. Administrative dialing

    d. Call processing

15. The basic system component of the DEFINITY Server digital switch is the:

    a. Data port

    b. Voice port

    c. Port circuit

    d. PN

16. A sales department wants to do call costing. Which of the following applications could be used to best fit their needs?

    a. ACD

    b. PBX

    c. SMDR

    d. Call vectoring

17. A company has 350 users of their telephone system. Several of their departments handle technical support and sales. They have been getting complaints about no answers at certain times of the day. They request an automated attendant and would like to track all incoming and outgoing calls. Which applications are necessary to provide a solution?
    (Choose three.)

    a. Voice mail

    b. SMDR

    c. ACD

    d. Key system

    e. Wireless system

    f. Unified messaging

18. The PBX feature that allows a user to transfer incoming telephone calls to another telephone extension on a PBX system is:

    a. Call pickup

    b. Call forwarding

    c. Call waiting

    d. DAU

19. The central interface between a PPN and EPN is the:

    a. Data ports network

    b. Expansion port interface

    c. CSS

    d. PN

20. A company would like all of their calls answered by an automated attendant. The automated attendant should provide the caller several options for different departments within the organization. Which of the following products is necessary for this solution?

    a. Call vectoring

    b. Call accounting

    c. Voice mail

    d. Call management system

21. A company has three locations, one main office and two remote offices. They need to have both data and voice transferred between each remote office and the main office. They would like to track all calls made by employees. They would also like all incoming calls to be answered by a receptionist at the main office, who is backed by an automated attendant. Which of the following design scenarios meets their needs?

    a. PBX with three nodes, call accounting, centralized voice mail, T1s between each node

    b. PBX, call accounting, voice mail system

    c. PBX with three nodes, call accounting, T1s between each node, ACD with ARS

    d. PBX, ACD, SMDR only

22. After answering a call, an employee can transfer the call to another extension through which of the following features?

    a. Call forwarding

    b. Call pickup

    c. Speed dialing

    d. Call transfer

23. The devices used to easily connect and reconfigure one physical circuit to another are referred to as:

    a. RJ11

    b. Consoles

    c. Cross-connects

    d. PSTN

24. An ACD connects incoming calls to:

    a. Hunt groups

    b. Tie lines

    c. PBX software

    d. Processor bus

25. Which of the following directly impacts the hardware capacity of a PBX system?

    a. Trunk interfaces

    b. Available slots

    c. Station interfaces

    d. System adjuncts

26. Incoming calls are connected directly to employees of an organization by means of:

    a. FX

    b. DISA

    c. DID

    d. Attendant console

27. Which of the following is not a component found in a PBX system?

    a. Trunks

    b. Attendant console

    c. Administrative terminal

    d. DID

28. The system or PBX feature that routes calls to telephone support personnel is referred to as:

   a. UPS

   b. DSS

   c. ACD

   d. SMDR

29. SLC circuits are used to:

   a. Reduce the number of physical wires

   b. Increase the availability of tie lines

   c. Decrease the chances of power failure and loss of telephone service

   d. Reduce the size of equipment cabinets

30. Which of the following should be considered when increasing the capacity of a PBX system to handle more traffic? (Choose two.)

   a. Grounding and cabling

   b. Busy-hour statistics

   c. IP address availability

   d. Carrier capacity

   e. Current load on existing trunks

31. When more trunks and lines are needed for connectivity to and from a digital switch such as the DEFINITY Server, which of the following components must be added?

   a. Data ports

   b. Expansion ports

   c. Port circuits

   d. PNs

32. The feature of a PBX system that creates a log of all incoming and outgoing calls is called:

    a. UPS

    b. DSS

    c. Call accounting

    d. SMDR

33. What component is necessary to translate TDM voice communications into IP and vise-versa?

    a. A media server

    b. A transceiver

    c. A modem

    d. A media gateway

34. What types of services does a media server provide? (Choose two.)

    a. Call processing

    b. Protocol translation

    c. Feature functionality

    d. Email and voice mail integration

35. A company is upgrading a branch office location. What questions should you ask them to try and determine their availability needs? (Choose three.)

    a. Is it necessary to have voice mail functionality if the WAN link is down?

    b. Is it necessary to be able to make internal calls if the WAN link is down?

    c. Is it necessary to be able to make external calls if the WAN link is down?

    d. Who are the key players in this project?

Module 3
Point-to-Point
Telecommunications Protocols

In this module, we describe each point-to-point transmission service available from telecommunications providers. These services are essential for creating private voice networks or wide area data networks. Point-to-point connectivity is primarily concerned with Physical Layer protocols, which transfer information across a physical medium such as wire or optical fiber.

Each lesson in Module 3 focuses on one point-to-point transmission option. Together, these services form a gradual progression, from slow analog dial-up lines to super high-speed optical backbone systems. The point-to-point protocols presented in this module will be discussed in future modules as well, as we look at higher-layer protocols such as Integrated Services Digital Network (ISDN) and frame relay, and how they use these underlying Physical Layer services.

Lessons

1. Physical and Logical Circuits
2. Summary of Data Rates
3. Dial-Up and Leased Lines
4. T-Carriers
5. Asymmetric Digital Subscriber Line
6. Cable Modems
7. Synchronous Optical Network

Terms

Attenuation—The weakening of a signal over distance is referred to as attenuation.

Bridge—A bridge is a device that operates at the Data Link Layer of the OSI model. A bridge can connect several LANs or LAN segments. It can connect LANs of the same media access type such as two Token Ring segments, or different LANs such as Ethernet and Token Ring.

Cable Vault—A cable vault is a below ground box, roughly 4 x 8 feet, that provides maintenance access to buried cable. It is also called a "maintenance hole."

Digital Access Cross-Connect Switch (DACS)—DACS is a connection system that establishes semipermanent (not switched) paths for voice or data signals. All physical wires are attached to a DACS once, then electronic connections between them are made by entering instructions.

E1—E standards are the European standards that are similar to the North American T-carrier standards. E1 is similar to T1.

Frame Relay—Frame relay is a packet-forwarding WAN protocol that normally operates at speeds of 56 Kbps to 1.5 Mbps.

High Speed Serial Interface (HSSI)—HSSI is a serial data interface that can operate at speeds up to 52 Mbps.

Hybrid Fiber-Coax (HFC)—HFC is a network design method, common in the cable television industry, that combines optical fiber and coaxial cable into a single network. Fiber optic cables run from a central site to neighborhood hubs. From the hubs, coaxial cable serves individual homes.

Interference—Any energy that interferes with the clear reception of a signal is referred to as interference. For example, if one person is speaking, the sound of a second person's voice interferes with the first. See noise.

International Telecommunication Union-Telecommunications Standardization Sector (ITU-T)—ITU-T is an intergovernmental organization that develops and adopts international telecommunications standards and treaties. ITU was founded in 1865 and became a United Nations agency in 1947.

Local Loop—The pair of copper wires that connects a customer's telephone to the LEC's CO switching system is referred to as the local loop.

Management Information Systems (MIS)—MIS is the traditional name of the department responsible for a company network or computing infrastructure.

Multiplexer (MUX)—MUX refers to computer equipment that allows multiple signals to travel over a single channel. Multiple signals are fed into a MUX and combined to form one output stream.

Noise—Any undesired signal or signal distortion is referred to as noise. Noise is often caused by some kind of interference. See interference.

Optical Carrier (OC)—OC is one of the optical signal standards defined by the SONET digital signal hierarchy. The basic building block of SONET is the STS-1 51.84-Mbps signal, chosen to accommodate a DS3 signal. The hierarchy is defined up to STS-48 (48 STS-1 channels), for a total of 2,488.32 Mbps capable of carrying 32,256 voice circuits. The STS designation refers to the interface for electrical signals. The corresponding optical signal standards are designated OC-1, OC-2, etc.

Permanent Virtual Circuit (PVC)—A PVC is a connection across a frame relay network, or cell-switching network such as ATM. A PVC behaves like a dedicated line between source and destination end-points. When activated, a PVC will always establish a path between these two end points.

Phase—A phase is a description of one wave's position relative to another at a particular point in time. Phase differences are measured in degrees from 0 to 180.

Router—A router is a Layer 3 device, with several ports that can each connect to a network or another router. A router examines the logical network address of each packet, then uses its internal routing table to forward the packet to the routing port associated with the best path to the packet's destination. If the packet is addressed to a network that is not connected to the router, the router will forward the packet to another router that is closer to the final destination. Each router, in turn, evaluates each packet, then either delivers the packet or forwards it to another router.

Span—A digital connection between a CO and terminal switch, such as a PBX, is referred to as a span.

Switched Multimegabit Data Service (SMDS)—SMDS is a connectionless service used to connect LANs, MANs, and WANs at rates up to 45 Mbps. SMDS is cell-oriented and uses the same format as the ITU-T B-ISDN standards. The internal SMDS protocols are called SIP-1, SIP-2, and SIP-3. They are a subset of the IEEE 802.6 standard for MANs, also known as DQDB.

Switched Virtual Circuit (SVC)—An SVC is a temporary connection established through a switched network. During data transmission, an SVC behaves like a wire between the sender and receiver. ATM VC and telephone connections are both examples of SVCs.

T1, T3—T1 and T3 are two services of a hierarchical system for multiplexing digitized voice signals. The first T-carrier was installed in 1962 by the Bell System. The T-carrier family of systems now includes T1, T1C, T1D, T2, T3, and T4 (and their European counterparts E1, E2, etc.). T1 and its successors were designed to multiplex voice communications. Therefore, T1 was designed such that each channel carries a digitized representation of an analog signal that has a bandwidth of 4,000 Hz. It turns out that 64 Kbps is required to digitize a 4,000-Hz voice signal. Current digitization technology has reduced that requirement to 32 Kbps or less; however, a T-carrier channel is still 64 Kbps. A T1 line offers bandwidth of 1.544 Mbps; a T3 offers 44.736 Mbps.

T-span—A T-carrier that connects a PBX to a CO is referred to as a T-span.

X.25—X.25 is a connectionless packet-switching network, public or private, typically built upon leased lines from public telephone networks. In the United States, X.25 is offered by most carriers. The X.25 interface lies at OSI Layer 3, rather than Layer 1. X.25 defines its own three-layer protocol stack, and provides data rates only up to 56 Kbps.

Lesson 1—Physical and Logical Circuits

This module reviews the physical foundation that makes up virtually all networks in use today. This lesson will familiarize you with circuit types used for communication over wide area network (WAN) facilities.

Objectives

At the end of this lesson you will be able to:

- Describe the concept of a virtual circuit

- Describe the difference between a switched virtual circuit (SVC) and permanent virtual circuit (PVC)

- Understand basic concepts behind WAN information transfer

- Describe the layers where data terminal equipment (DTE) and data communications equipment (DCE) function in a WAN

 **Key Point** | *A virtual circuit has the appearance of a single physical circuit.*

Circuits and Virtual Circuits

A circuit is the physical connection between two communicating devices. A circuit is also referred to as a channel. Physical circuits can be classified into two broad categories:

- Exclusive-use physical circuit

- Shared-use physical circuit

An exclusive-use physical circuit is one in which a computer and attached devices do not share the circuit with any other device. Examples are the physical circuit connecting a monitor to a video card and the circuit connecting a printer to the parallel port. This is shown on the Exclusive-Use Physical Circuit Diagram.

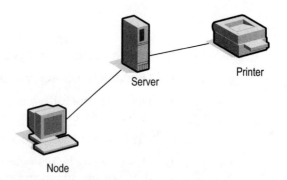

Exclusive-Use Physical Circuit

A shared-use physical circuit is one in which multiple devices share the same physical media. Access to the shared media depends on the media access protocol used (such as Token Ring or Ethernet). The Shared-Use Physical Circuit Diagram demonstrates this arrangement, also called a bus.

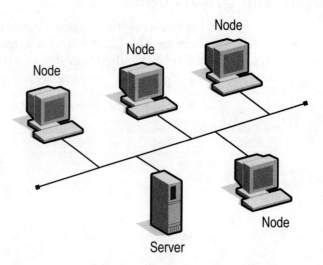

Shared-Use Physical Circuit

A virtual circuit is a communications path that appears to be a single circuit, even though the data may take varying routes between the source and destination nodes. A virtual circuit is not a physical circuit; a virtual circuit is a logical connection between devices configured over one or more physical circuits.

A Transmission Control Protocol/Internet Protocol (TCP/IP) connection between Internet devices is an example of a virtual circuit; the devices establish a connection for the duration of the conversation, then disconnect when the conversation ends. This connection traverses many different physical circuits, including a telephone line or digital connection to the telephone company central office (CO), the CO's connection to the Internet, and the Internet's connection to the target server's network. These physical circuits remain in place once the conversation terminates (the telephone company doesn't remove your telephone line when you disconnect from the Internet); only the logical circuit terminates.

This concept has its roots in X.25 (an International Telecommunications Union-Telecommunication Standardization Sector [ITU-T]) packet switching. The Virtual Circuit Diagram illustrates this concept.

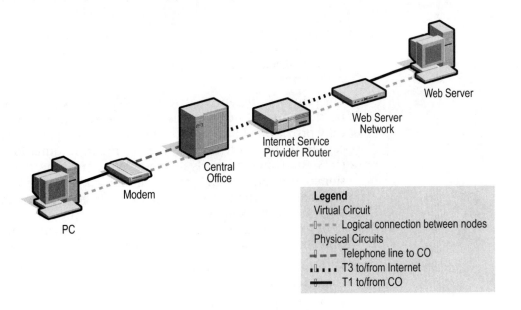

Virtual Circuit

There are two types of virtual circuits:

- A PVC behaves like a dedicated connection between source and destination endpoints. A PVC will always establish a pre-defined path between these two endpoints.

- An SVC is analogous to a public-switched telephone service, in that calls can be made dynamically between a source endpoint and any destination endpoint in a network. The end devices establish a connection on demand, and disconnect when communications cease.

PVCs and SVCs

The choice of whether to use a PVC or SVC varies between networks depending on traffic volumes, traffic patterns, degree of connectivity, types of applications, and other parameters. The PVCs and SVCs Table presents some differences between the two types of virtual circuits. Networks can have a mix of PVCs and SVCs.

PVCs and SVCs

Permanent Virtual Circuit	Switched Virtual Circuit
Statically defined at configuration	Dynamically established when there is information to send
Connection always configured regardless of whether there is information to send	Connection released when there is no more information to send

The primary difference between a PVC and SVC occurs when connections are defined and resources are allocated. PVCs are typically provisioned by a network operator, whether the operator is a carrier (public services) or Management Information Systems (MIS) staff member (private networks). After the PVC is provisioned, the connection is available for use at all times unless there is a change in the service or a service outage. On the other hand, SVCs are established by the end user, not by the network operator. Prior to each use, an SVC connection is established to the destination end user. The connection is cleared after each use.

SVC applications are ideal for networks that have the following characteristics:

- Highly meshed connectivity

- Intermittent applications

- Remote site access

Highly meshed connectivity refers to large networks that need any-to-any connectivity; not to be confused with meshed networks that consist of multiple nodes connected by many dedicated physical connections. Multiple virtual circuits can traverse the same physical connections, connecting multiple sites simultaneously.

When many networked locations must communicate with each other, SVCs may offer the best solution. The advantages of SVCs are magnified as the number of locations and degree of connectivity requirements increase. Highly meshed networks are becoming more common as more and more companies use intranets. It is conceivable that all end users will have their own World Wide Web (Web) pages within an organization. This will increase the amount of peer-to-peer intracompany traffic. Additionally, a highly meshed network can offer a cost-effective solution for occasional intercompany connections to suppliers, partners, and even customers, provided they all subscribe to the same service.

A network that has highly intermittent applications usually generates traffic that is unpredictable and short in duration, such as electronic mail (e-mail) traffic. Because SVCs only consume network bandwidth when there is information to send, they are a good solution for short-duration applications.

Initially, some small office and telecommuter locations may not be able to justify the cost of PVCs for all locations to which they need connectivity because of low traffic volumes and intermittent use. These locations can start with SVCs and gradually migrate to PVCs as traffic volumes increase. A hybrid PVC/SVC solution can be implemented as well. PVCs are often established between locations that require frequent exchange of information, and SVCs are established at locations that only need occasional interaction.

SVC Information Transfer

The transfer of information across an SVC consists of three primary events as follows:

1. Call setup

2. Data transfer

3. Call release

The SVC Sequence Diagram illustrates these steps.

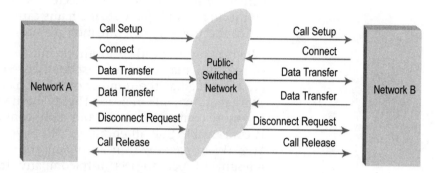

SVC Sequence

When information needs to be sent across an SVC, a setup message is sent to a destination network. After the setup phase is complete and the connection has been established, information can be sent back and forth across the network.

The call release phase consists of a disconnect message sent from the source to the destination. The connection to the network is released at both networks, and a circuit no longer exists between the two networks.

Connecting to WAN Circuits

WAN circuits are used to move different types of information across a telecommunications network. There are different ways to connect to a WAN, which is the subject of this section.

DCE and DTE

The traditional telephony name for equipment that requests services is data communications equipment (DCE) or data circuit-terminating equipment (DTE). DCE can include such items as:

- Modem, if the transmission channel is an analog channel (that is, at least the local loop is an analog channel)

- Digital-encoding device, if the transmission channel is a digital channel

DCE devices are Open Systems Interconnection (OSI) model Layer 1 devices responsible for properly formatting the electrical signals on a physical link, and performing signal clocking and synchronization.

DTE executes Layer 2 and higher processes. DTE is often a computer, but may also be a terminal or other specialized equipment, such as an automated teller machine.

The DCE and DTE Diagram illustrates the difference between the two types of equipment.

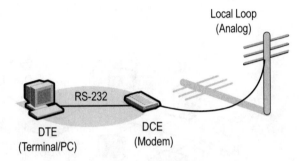

DCE and DTE

DTE-to-DCE Protocols

Both the DTE and DCE interfaces lie at Layer 1 of the OSI model. Many different standards exist for the DTE/DCE interface. Two important standards are the Electronic Industries Association (EIA) RS-232D standard and the wideband ITU-T V.35 standard.

Both standards specify the Physical Layer (mechanical and electrical) properties of the interface and its procedural characteristics. RS-232D uses a 25-pin connector, called a DB25, for the cable that runs between the DTE and DCE (as well as a 9-pin variant); V.35 uses a 34-pin V.35 connector. There are still additional protocols to be considered: those between DTEs (through DCEs), and those between DCEs themselves.

Layer Upon Layer

In most environments, protocols are grouped according to their association with a local area network (LAN) or WAN. For example, communication might take place between a computer on one LAN with a computer on another LAN using a Layer 3 protocol, such as the TCP/IP stack. TCP/IP packets can be transported across a WAN using a Layer 2 WAN protocol such as frame relay over Layer 1 T1 circuits (discussed later).

The type of circuit and connection used between LANs may be either PVC or SVC, depending on the protocols used. For example, if we have two LANs we want to connect directly, we might use a PVC as shown on the Permanent Virtual Circuit Diagram. This PVC may traverse a digital circuit that only executes Physical Layer protocols, such as Digital Data Service (DDS) or T1, and provides only point-to-point connectivity.

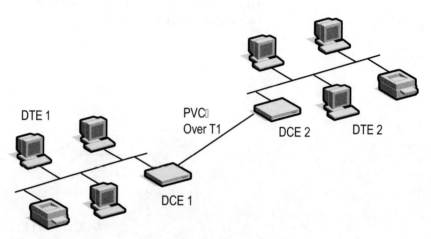

DTE 1

PVC
Over T1

DCE 2

DTE 2

DCE 1

Permanent Virtual Circuit

Information generated by DTE 1 travels through DCE 1, across the PVC to DCE 2. This device then transfers the information across the LAN to the destination DTE (DTE 2).

The Switched Virtual Circuit Diagram illustrates connectivity between two DTEs attached to an SVC. In this diagram three, and potentially many more, LANs are connected to a telecommunications service that provides switched connectivity. If DTE 3 were to send data to DTE 4, the DCEs that attach each of these devices to the network would first establish a connection between themselves. After this virtual circuit is established, the DTEs could send information to each other. After communication between DTEs is complete, the virtual circuit is terminated.

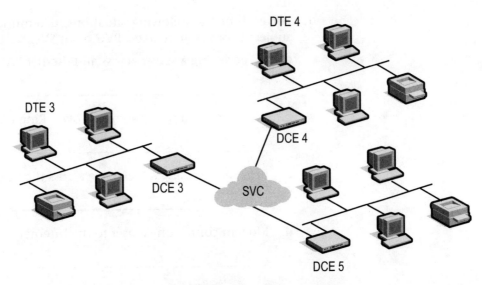

Switched Virtual Circuit

It should be noted that a computer is rarely referred to as a DTE; however, in telecommunications terminology, the term is still widely used to group together devices that access a WAN. Also note that DCEs are not necessarily devices that connect to "cabled" networks. These devices are capable of sending information by the means of other methods, such as microwaves and satellite technology.

Activities

1. For each of the following situations, determine whether the subject behaves more like a PVC or an SVC:

   a. PC accessing a server across an Ethernet LAN

   _____

   b. PC accessing a server across a Token Ring LAN

   _____

   c. Browser accessing a Web page across the Internet

   _____

   d. Modem connecting a user to the Internet

   _____

   e. Two networks tied together with a T1 circuit

   _____

2. List the applicable DTE and DCE interfaces for the following connections:

   a. PC to external modem

   _____

   _____

   b. Terminal connecting to a PVC by means of a digital encoding device

   _____

   _____

   c. Computer connecting to an SVC by means of a modem

   _____

   _____

   d. Terminal connecting to another terminal across a dial-up connection by means of a modem

   _____

   _____

Extended Activities

1. Find out how X.25 packet-switching networks work, and how they illustrate the concept of an SVC.

2. What types of networks use PVCs? How do users of these services benefit over those who use SVCs? How do users of SVCs benefit over users of PVCs?

3. Using the Web, research other Physical Layer standards for DTE-to-DCE connectivity. Include in your research the following standards:

   a. RS-422

   b. RS-423

   c. RS-449

Lesson 2—Summary of Data Rates

This lesson summarizes the most popular Physical Layer protocols that provide connectivity to WAN and metropolitan area network (MAN) environments. We begin this lesson by looking at low-speed technologies such as dial-up and leased lines, and we progress to higher-speed technologies such as T3 and Optical Carrier (OC)-3. We will look at applications that might use each of these options, and the associated performance that each application may require.

Objectives

At the end of this lesson you will be able to:

- Name the low- and high-speed options for Physical Layer MAN/WAN connectivity

- Determine which technology is most appropriate for a given business application

 **Key Point**

WAN Physical Layer protocols vary widely in speed and cost.

Point-to-Point Links

Point-to-point links establish a physical connection between local and remote stations. These links come in a variety of data rates, and as speed and capability increase, cost increases as well. Because a point-to-point link provides dedicated bandwidth for the life of the circuit, the cost of moving data this way is usually much higher than with switched services. In addition, when constructing a MAN or WAN using point-to-point links, we must buy these dedicated facilities for each line of communication we want to establish. Thus, to establish all-to-all connectivity, the number of links increases rapidly with the number of nodes: 3 links for 3 nodes, 10 links for 5 nodes, and so on.

Switched services let us establish the necessary number of dedicated links as virtual circuits over a shared communications service. Switched services such as Asynchronous Transfer Mode (ATM) and frame relay continue to be a primary alternative when connecting remote networks.

Analog Lines

The bottom rung of the point-to-point ladder is the analog connection, using modems to carry data over leased or switched lines. Local exchange carriers (LECs) offer these traditional telecommunications services over existing telephone company voice network facilities and the copper local loop between a customer and CO. Leased lines are full-time connections between two specified locations; switched lines are regular telephone lines, often called "plain old telephone service" (POTS).

Modems became extremely popular in the first half of the 1990s; however, increased broadband technology availability limits their use in special applications and to those unfortunate enough to have no other option for Internet connectivity. The fastest modems today adhere to the ITU-T V.92 recommended standards, operating at a compressed rate of approximately 53 kilobits per second (Kbps) downstream and 48 Kbps upstream. These rates, although theoretically attainable, are often impractical given the signal-to-noise ratio on unconditioned voice-grade telephone circuits. Additionally, when a modem call originates on a PBX, such as when calling from a hotel room, the additional analog-to-digital conversion step limits bandwidth in both directions to no more than 28.8 Kbps.

DDS

Digital Data Service (DDS) is also called Dataphone Digital Service. It provides speeds ranging from 2.4 to 56 Kbps. DDS lines are full-time leased connections between two specified locations, and support a fixed bandwidth. They are usually used to construct private digital networks. Connections are made to DDS using a special device called a data service unit/channel service unit (DSU/CSU). A DSU/CSU functions on a digital line as a modem functions on an analog line.

SW56

Next on the point-to-point ladder, after DDS, is Switched-56 (SW56) service, which enables dial-up digital connections to any other SW56 subscriber in the country. SW56 service uses a DSU/CSU just as a leased-line DDS; however, it includes a dialing pad for entering the telephone number of the remote SW56 system.

High-Speed Services

Finally, at the top of the point-to-point ladder are the truly high-speed digital services, including:

- Fractional T1 (FT1)
- T1
- T3
- Synchronous Optical Network (SONET)

259

Various Data Rates and Associated Applications

The Physical Layer Technologies Table presents key Physical Layer technologies most often used for connection to a WAN. Associated data rates, physical media, and applications are also listed. Technology choices are based on need and economics. Each of these is discussed in detail in the lessons that follow.

Physical Layer Technologies

Technology	Data Rate	Physical Media	Application
Dial-up	14.4 to 56 Kbps	Low-Grade Twisted Pair	Home Office Connectivity to Office and Internet
DDS Leased Line	56 Kbps	Low-Grade Twisted Pair	Small Business Low-Speed Access Office-to-Office Connectivity Internet Connectivity
Switched-56	56 Kbps	Low-Grade Twisted Pair	Small Business Low-Speed Access Office-to-Office Connectivity Internet Connectivity Link Backup
Fractional T1	64 to 768 Kbps	Low-Grade Twisted Pair	Small to Medium Business Moderate-Level Speed Internet Access
Satellite (DirecPC)	400 Kbps downstream 33.6 Kbps upstream	Radio Spectrum	Small Business with Moderate-Level Speed Internet Access
T1	64 Kbps to 1.544 Mbps	Low-Grade Twisted Pair Optical Fiber Microwave	Medium Business Internet Access Point-to-Point LAN Connectivity
E1	64 Kbps to 2.048 Mbps	Low-Grade Twisted Pair Optical Fiber Microwave	Medium Business Internet Access Point-to-Point LAN Connectivity
ADSL	128 to 8 Mbps	Low-Grade Twisted Pair	Medium Business High-Speed Home Internet Access
Cable Modem	512 Kbps to 52 Mbps	Coaxial Cable	Medium Business High-Speed Home Internet Access

Physical Layer Technologies (Continued)

Technology	Data Rate	Physical Media	Application
E3	34.368 Mbps	Twisted Pair Fiber Optic Cable Microwave	Large Business Internet Access ISP Backbone Access
T3	45 Mbps	Twisted Pair Fiber Optic Cable Microwave	Large Business Internet Access ISP Backbone Access
OC-1	51.48 Mbps	Fiber Optic Cable	Backbone, Campus Internet to ISP
OC-3	155.52 Mbps	Fiber Optic Cable	Large Company Backbone Internet Backbone Connectivity
OC-24	1.24 Gbps	Fiber Optic Cable	Large Company Backbone Internet Backbone Connectivity
OC-48	2.5 Gbps	Fiber Optic Cable	Large Company Backbone Internet Backbone Connectivity
OC-192	10 Gbps	Fiber Optic Cable	Internet Backbone Connectivity

Bandwidth

Bandwidth is the difference between the highest and lowest frequencies that can be transmitted across a transmission line or through a network. It is measured in hertz (Hz) for analog networks, and bits per second (bps) for digital networks.

Different types of applications require different bandwidths for effective use, as shown in the Application Bandwidth table.

Application Bandwidth

Application	Required Bandwidth
Personal Computer (PC) Communications	300 bps to 56 Kbps
Digital Audio	128 Kbps to 1 megabit per second (Mbps)
Compressed Video	128 Kbps to 2 Mbps
Document Imaging	10 to 100 Mbps
Full-motion Video	250 Kbps to 2 gigabits per second (Gbps)

Some types of physical transmission media, such as coaxial cable, can carry multiple simultaneous signals by assigning each signal to a range of frequencies, just as a radio does. Thus, the greater the range of frequencies a medium can handle, the greater its information-carrying capacity. For example, most analog modems transmit data within a 300- to 3,000-Hz frequency range in the middle of a bandwidth.

Although signal characteristics are usually optimal in the middle of a bandwidth, transmission limited to the middle of the band restricts the amount of bandwidth available for data. To compensate for this factor, conventional modems use sophisticated, multiple-bit encoding algorithms to squeeze as much data as possible over one channel in each direction. A disadvantage of this solution, however, is an increase in the amount of data lost during line hits or other error-inducing conditions on the transmission medium. One goal of much of modem design work is to minimize data losses while transferring larger amounts of data.

Activities

1. For each technology listed below, find router products that support each of the physical interfaces:

    a. SW56

    _____

    b. T1

    _____

    c. ADSL

    _____

    d. OC-1

    _____

    e. OC-3

    _____

Extended Activities

1. Research the availability of each of these technologies in your area and determine which ones you can use.

2. Research the cost of the various technologies listed above, and the pricing structure associated with each as it relates to short-distance communication and long-distance communication.

3. For each of the technologies listed above, determine how long it would take to get a 5 MB file across a network using this technology.

   a. SW56

   _____

   b. T1

   _____

   c. ADSL

   _____

   d. OC-1

   _____

   e. OC-3

   _____

Lesson 3—Dial-Up and Leased Lines

Standard telephone lines can be used for transmission of digital and analog information. This lesson looks at the predominant means of transmitting information across standard telephone lines: dial-up and leased-line technologies.

Objectives

At the end of this lesson you will be able to:

- Describe the characteristics of dial-up connectivity

- List the advantages and disadvantages of dial-up vs. leased-line technologies

 **Key Point**

Dial-up connectivity uses the switched network provided by the telephone company.

Dial-up Connections

A dial-up line is a circuit between two nodes, established across the switched telephone network as illustrated on the Switched Line Diagram. Dial-up lines provide the following characteristics:

- 2.4 to 56-Kbps transfer rates

- Any-to-any connectivity (one at a time)

- Compatible modems at each end

- Call initialization required before transmission

- Inexpensive

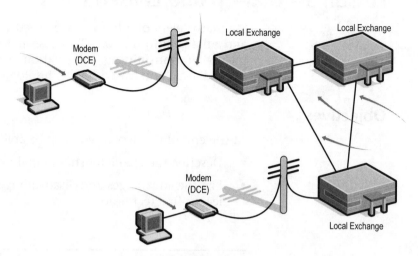

Switched Line

Leased Lines

Leased lines are set up on a permanent basis by a LEC or long distance carrier. Leased lines are most appropriately used when the most important traffic requirements are steady and uninterrupted service. The most popular type of leased line is T1. T1 equipment is readily available to carry both voice and data traffic in increments of 56 or 64 Kbps.

Advantages of leased lines over private lines are:

- Information security
- Constant quality of service (QoS)
- Circuit control

Disadvantages of leased lines over private lines are:

- Expense is constant, even when usage is not.
- As the number of connections increases, so do equipment needs and costs.

DDS

Digital signaling offers much more bandwidth than analog signaling, and at higher reliability. By eliminating the conversion of digital data to an audio signal and back again, a digital signaling system eliminates many of the problems modems must deal with: audio noise, phase and frequency shift, clock synchronization, variable line quality, and signal attenuation. The electronics for attaching DTE devices to a digital link are also much less complex, which in the end results in much less expense for equivalent bandwidth.

DDS links are leased, permanent connections, running at fixed rates of 2.4, 4.8, 9.6, 19.2, or 56 Kbps. A DSU/CSU device at each end provides the interface between a two-wire DDS line and traditional computer interfaces such as RS-232. A typical LAN interconnection uses two DDS-compatible bridges and external DSU/CSUs, as shown on the DDS Connectivity Diagram.

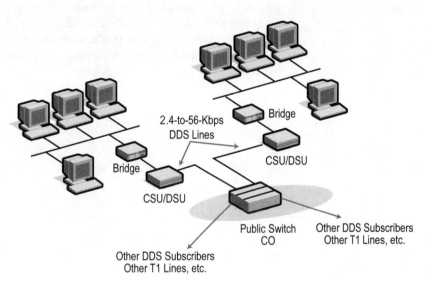

DDS Connectivity

Inside the CO, a DDS line is merged into the regular flow of traffic on T1 and T3 carrier facilities, which route it to its destination. The DDS route is established when the service is purchased, and bandwidth on the necessary trunk carriers is carved out at that time. We pay a fixed monthly fee for DDS, plus mileage charges based on the interoffice distance traversed over telephone company trunks. The requested data rate determines the fixed monthly fee.

DDS has physical limitations primarily related to the distance between a DSU/CSU and the serving CO. DDS works reliably when the route distance between a subscriber and a CO is less than 30,000 feet (local loop cable length). An office only 1 or 2 miles from a CO may nevertheless have 20,000 feet or more of intervening cable, due to the circuitous routes local loops often take in metropolitan areas.

Most telephone companies use a designated line when providing DDS; telephone company engineers trace the shortest possible route over existing copper facilities to get from the CO to a subscriber's location. Telephone company field technicians then visit cable vaults along the route to make the necessary physical connections establishing the designated route.

Activities

1. Compare and contrast leased lines vs. dial-up circuits.

   _____

   _____

2. On the diagram below, list the analog and digital parts of the circuit.

   _____

   _____

   _____

   _____

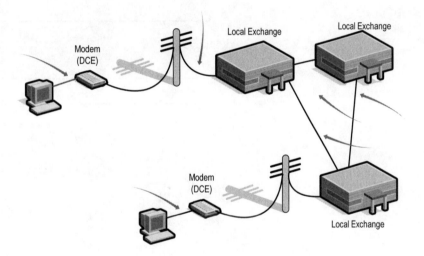

Extended Activity

1. Using the Web, find information on the following types of products:

    a. CSU

    _____

    _____

    b. DSU

    _____

    _____

    c. DDS

    _____

    _____

    d. DDS II

    _____

    _____

Lesson 4—T-Carriers

T-carriers (E-carriers in Europe) have existed in the telecommunications industry for quite some time. They are still widely used digital technologies for business voice and data connectivity, and are commonly used to connect private branch exchange (PBX) systems to a telephone company CO.

A digital connection between a CO and terminal switch, such as a PBX, is commonly called a span. Thus, T-carrier services such as T1 and T3 are commonly called T-spans.

Objectives

At the end of this lesson you will be able to:

- Describe the differences between T1, FT1, and Digital Signal Level 3 (DS3) systems

- List applications that use T1, FT1, and DS3 technologies

 **Key Point**

T-carriers are a primary means of point-to-point network connectivity.

Digitizing Voice

With the exception of the analog telephone signals generated by handset microphones and reproduced in handset speakers, today's telephone network is mostly digital. Telephone switches are complex computers that execute specialized routing protocols to establish voice connections. Control signals, such as a dial tone, busy signal, and ring, are enabled by an out-of-band digital signaling network.

Converting the telephone network from analog to digital became possible in the late 1950s, when solid-state electronics became available. The conversion of the U.S. telephone network, then serving 180 million telephones, was essentially complete less than 20 years after the basic technology became available.

Digital voice transmission offers several advantages:

- Digital signals are less susceptible to interference; it is easier to distinguish noise from signal.

- Binary patterns can be reproduced exactly, and "cleaned up," when they pass through switches, multiplexers (MUXs), or repeaters.

- Digital voice signals can be multiplexed with other digital data.

T-Carriers

In 1962, the Bell System installed the first time-division multiplexing (TDM) system for multiplexing digitized voice signals. Frequency-division multiplexing (FDM) had previously been developed to multiplex analog voice signals. However, the new "T-carriers" provided much better transmission quality.

The T-carrier family of systems now includes T1, T1C, T1D, T2, T3, and T4 (and their European counterparts E1, E2, etc.). Because smaller T-carriers are multiplexed into larger ones, the T-carrier system is also known as the North American Digital Signal Hierarchy. Thus, T1 is equivalent to DS1, T3 to DS3, and so on. Each of these levels is also called "American Standard" T-carriers, to distinguish them from the different data rates of the European versions. T-carrier data rates are shown in the T-Carrier Rates Table.

T-Carrier Rates

Standard	Line Type	Number of Voice Circuits	Bit Rate
North America			
DS0	N/A	1	64 Kbps
DS1	T1	24	1.544 Mbps
DS1C	T1C/D	48	3.152 Mbps
DS2	T2	96	6.312 Mbps
DS3	T3	672	44.736 Mbps
DS4	T4	4,032	274.176 Mbps

T-Carrier Rates (Continued)

Standard	Line Type	Number of Voice Circuits	Bit Rate
Europe			
E1	M1	30	2.048 Mbps
E2	M2	120	8.448 Mbps
E3	M3	480	34.368 Mbps
E4	M4	1,920	139.264 Mbps
E5	M5	7,680	565.148 Mbps
Japan			
1	F-1	24	1.544 Mbps
2	F-6M	96	6.312 Mbps
3	F-32M	480	34.064 Mbps
4	F-100M	1,440	97.728 Mbps
5	F-400M	5,760	397.20 Mbps
6	F-4.6G	23,040	1,588.80 Mbps

T1 and its successors were designed to multiplex voice communications. Therefore, each channel of a T1 line was designed to carry a digitized representation of an analog signal that has a bandwidth of 4,000 Hz. A T-carrier channel, DS0, requires 64 Kbps to digitize a 4,000-Hz voice signal, because the most accurate digital representation of an analog signal occurs when the signal is sampled at twice its bandwidth. When a DS0 is sampled, 8 bits represent each sampled amplitude. Eight bits times 8,000 samples per second equals 64,000 bps.

DS0 or Fractional T1

Leasing a T1 line means paying for the entire 1.544-Mbps bandwidth 24 hours a day, whether it is used or not. However, as we can see in the T-Carrier Rates Table, a T1 line is made up of 24 DS0 channels. Telephone companies offer a popular digital service, called FT1, that provides data rates from 64 Kbps to 1.544 Mbps in DS0 increments. For example, FT1 lets us lease any 64-Kbps submultiple of a T1 line. We might, for example, lease only six channels to obtain six 64-Kbps channels or an aggregate bandwidth of 384 Kbps.

FT1 is useful whenever the cost of a dedicated T1 is prohibitive, or a user only needs a portion of the normal T1 bandwidth (1.544 Mbps). FT1 is not as efficient or flexible as switched services, because we are paying to have a fraction of leased bandwidth available on a 24-hour basis. However, FT1 has an intrinsic feature not available with full T1 circuits: multiplexing DS0 channels outside our own enterprise T1 network.

Because we are not leasing an entire T1 circuit, we cannot dictate the location of the other end of the circuit. After all, we will be sharing the T1 with other customers. Each FT1 circuit terminates at a telephone company-managed digital access cross-connect switch (DACS). The far end of every FT1 circuit is embedded in the DACS, where the telephone company has established its own network of T1 interconnects.

Any two companies sharing the same DACS can switch among each other's DS0 channels, provided the telephone company has configured the two companies as interoperating organizations. This interoperability can be an advantage when a large central organization (e.g., government agency) needs to interoperate with many smaller organizations (e.g., contractors).

Due to the one-end nature of FT1, a separate FT1 circuit must be leased between each network node and the CO DACS. Contrast this with T1 circuits, where we need only lease one circuit between each pair of nodes. For this reason, as the number of T1 fractions increases, FT1 eventually becomes more expensive than T1. Typically, this occurs at approximately 75 percent of the full T1 bandwidth (18 channels).

T1

T1 circuits are dedicated services that connect networks or LANs over extended distances. The Sample T1 Configuration Diagram presents a typical configuration. This diagram illustrates how a T1 or T3 circuit could be used to connect two networks.

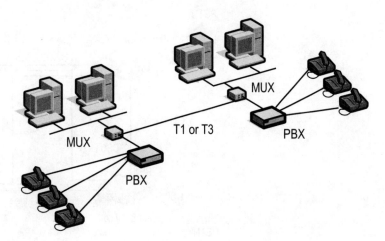

Sample T1 Configuration

As with a DDS circuit, we lease a T1 circuit between two locations. Unlike a DDS circuit, though, we have the ability to partition the bandwidth into up to 24 64-Kbps channels, with each configured to carry telephone calls or data traffic. Thus, a pair of T1 MUXs and a single T1 line can provide connectivity for both voice and data traffic across the same physical circuit.

The DS0 Network Diagram demonstrates this by showing a typical four-office telephone network built using individual DS0 lines. Several DS0 channels are allocated to carry "inside" telephone traffic between offices. These "tie lines" connect PBX systems at each location, forming a wide area private telephone network.

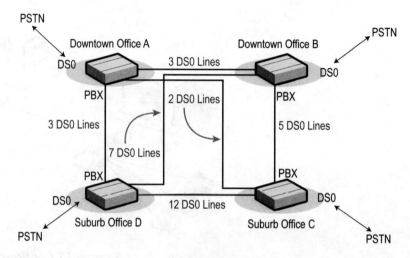

DS0 Network

Another group of DS0 channels is allotted to each office, to access the public-switched telephone network (PSTN) for "outside" calls. As traffic patterns change, lines must be moved or added between offices, a time-consuming and inconvenient task.

The T1 Network Diagram demonstrates a way to consolidate interoffice communications into a single backbone consisting of four T1 lines, capable of meeting changing traffic demands on the fly. Because T1 service includes the ability to switch DS0s within the local CO's network, any DS0 in one office can be switched to any DS0 in another office or to the PSTN. Data use of a T1 takes advantage of these abilities by means of a T1 MUX, a network terminating device that plays a role similar to that of an SW56 DSU/CSU. The difference is that the T1 MUX handles 24 DS0 channels instead of just 1 channel. A T1 MUX splices up to 24 64-Kbps DS0 channels into a single T1 1.544-Mbps bit pipe.

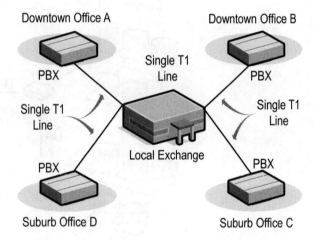

T1 Network

T1 and T3 circuits are useful WAN options because they offer adaptable bandwidth at essentially fixed costs within a metropolitan area. T1- and T3-capable routers and bridges typically support one or more T1 or T3 circuits, and automatically make connections with other routers or bridges in a network. The T1 WAN Diagram illustrates a typical T1-based WAN. As network designers, we program the router or bridge to specify the adjacent routers or bridges for each digital circuit. We can also specify bandwidth-on-demand parameters to change bandwidth in DS0 increments, similar to the technique used with SW56 lines.

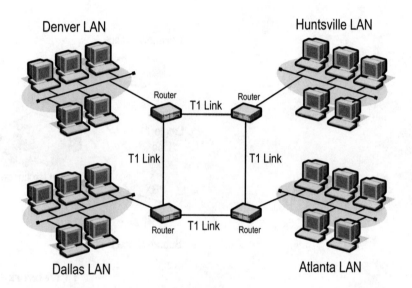

T1 WAN

T3 and North American Digital Signal Hierarchy

The North American Digital Signal Hierarchy is created using a series of TDMs, as shown on the North American Digital Signal Hierarchy Diagram. Four DS1 (T1) signals are fed into a DS2 MUX, which combines them to form one DS2 signal. Seven DS2 signals are combined to form one DS3 (T3) signal, and so forth.

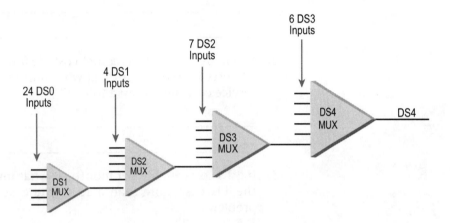

North American Digital Signal Hierarchy

DS3 (or T3) is another popular service offering for high-speed private lines. The T3 data rate is 44.736 Mbps.

Circuit Costs

Lower T1 circuit costs are enabling more and more companies to carry bandwidth-intensive applications, including video conferencing and other image-transfer programs. Business users who have lower-bandwidth, voice-grade digital circuits can upgrade to a T1 for only a short jump in cost.

This downward trend in T-carrier pricing is countered by the introduction of newer broadband Internet services, such as SDSL. As voice over Internet quality improves, T-carriers could lose even more ground to broadband technologies

Activities

1. What are the advantages and disadvantages of an FT1 over a T1 connection?

   _____

   _____

   _____

   _____

2. If one 64-Kbps DS0 channel costs $50 per month, and a full T1 costs $520 per month, at what multiple of DS0s would FT1 service cost more than a full T1?

   _____

   _____

3. If a T3 costs $4,200 per month, at what multiple of T1s would the T1s cost more than a T3 (using costs from the previous problem)?

   _____

   _____

Extended Activity

Using the Web, find information on at least three products that have T1 connectivity used for data communications. List the products and associated features.

_____

_____

_____

_____

_____

Lesson 5—Asymmetric Digital Subscriber Line

Asymmetric Digital Subscriber Line (ADSL) is intended for the last leg into a customer's premises, the local loop. As its name implies, ADSL transmits an asymmetric data stream, with much more going downstream to a subscriber and much less coming back.

Objectives

At the end of this lesson you will be able to:

- Explain why ADSL is asymmetrical

- Describe the advantages of ADSL over other local loop access methods

- Describe the main technical requirements for ADSL service

Key Point

ADSL provides more bandwidth downstream, toward a subscriber.

Digital Subscriber Lines and Internet Access

The local loop or "last mile" of the communications network deals with the Physical and Data Link Layers of the OSI model. Communications providers, such as cable companies, telephone companies, and satellite transmission companies, are currently investing billions of dollars in creating broadband infrastructure in the local loop. This section provides an overview of these developments.

Options for Local High-Speed Internet Access

The Internet Access Solutions Table summarizes mass-market solutions for Internet access currently used or being tested in some areas of the United States.

Internet Access Solutions

Variables	56-Kbps Modem	ISDN	ADSL Lite	RADSL	ADSL	Cable Modem	DirecPC Satellite
Speed to user	56 Kbps	128 Kbps	1.5 Mbps	7 Mbps	8 Mbps	30 Mbps	400 Kbps
Speed from user	33.6 Kbps	128 Kbps	128 Kbps	1 Mbps	1 Mbps	3 Mbps	None (must dial up)
Cost per month	$20	$60-100	$40-100	$40-200	$40-200	$30-60*	$20-130

*Monthly costs are estimates and include ISP-equivalent services (e.g., content, browser).

The 56-Kbps modem is the fastest dial-up solution available to consumers today. It is an inconvenient, "narrow band" solution, provided here for comparison with the broadband solutions.

ISDN is an end-to-end switched digital network that integrates enhanced voice and image features with high-speed data and text transfer. Built on top of standard unshielded twisted pair (UTP) telephone wire, ISDN provides two rates of service, basic and primary. The relevant version for the mass market is basic-rate ISDN, which provides three channels over one pair of twisted copper wires: two 64-Kbps bearer channels and one 16-Kbps data channel for signaling or packetized data. The two bearer channels can be bonded together to provide a total speed of 128 Kbps.

Although available since the early 1990s, ISDN has not caught on because of availability and price issues. (LECs have been reluctant to cannibalize their T1 business. ADSL and RADSL can deliver ISDN, although it takes away from the bandwidth for data.)

Cable modem is a technology for providing broadband Internet access over a cable television provider's hybrid fiber-coaxial (HFC) network. It is a broadcast technology analogous to an Ethernet LAN. Bandwidth is shared and packets move around in a store-and-forward scheme. The cable modem located in each subscriber's home filters out information not addressed to that particular subscriber. The remaining information is delivered to the subscriber's computer, by the means of a virtual point-to-point connection.

Cable modem downstream speeds can reach 36 Mbps or more, although all subscribers in the neighborhood usually share this bandwidth. Consequently, end-user connections usually run at less than 10 Mbps downstream and less than 1 Mbps upstream.

DSL

Digital Subscriber Line (DSL) is a modem technology that converts existing twisted pair telephone lines into access paths for multimedia and high-speed data communications, while simultaneously providing POTS. Developed in the 1980s to deliver video-on-demand (VOD) over telephone lines, DSL has the potential to deliver data at 160 times the speed of a 56-Kbps modem. This speed makes it a very attractive Internet access option for both homes and small businesses, and a very competitive service for telecommunications companies.

DSL achieves these high connection speeds by using the additional bandwidth that a copper pair can support. The ADSL Bandwidth Diagram shows how ADSL provides nearly 10,000 kilohertz (kHz) of bandwidth, compared to the 3 kHz of an analog POTS channel.

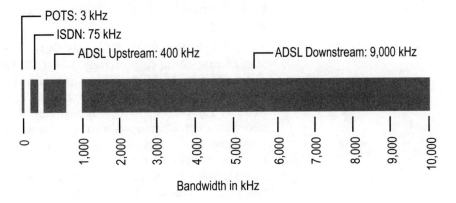

ADSL Bandwidth

DSL comes in two main formats: asymmetrical and symmetrical. Asymmetrical DSL provides faster download speeds than upload, which is the standard pattern for the Internet. With symmetrical DSL, data travels in both directions at the same speed.

Asymmetrical DSL technologies include:

- **Asymmetrical DSL (ADSL)**—ADSL offers differing upload and download speeds and can be configured to deliver up to six megabits of data per second (6000K) from the network to the customer—that is up to 120 times faster than dialup service and 100 times faster than ISDN. Upstream (from the user to the CO) data rates range from 128 Kbps to 640 Kbps. ADSL enables voice and high-speed data to be sent simultaneously over the existing telephone line. This type of DSL is the most predominant in commercial use for business and residential customers around the world. It is good for general Internet access and for applications where downstream speed is most important, such as streaming video. ITU-T Recommendation G.992.1 and ANSI Standard T1.413-1998 specify full rate ADSL.

- **ADSL Lite (Also known as G.lite ADSL, G.lite, or Universal ADSL)**—Developed to eliminate the splitters required in traditional ADSL installations, G.lite supports downstream speeds of up to 1.5 Mbps and upstream speeds of up to 512 Kbps. G.lite is more prone to noise since the voice and data are combined, but can be less expensive to install than ADSL.

- **Rate Adaptive DSL (RADSL)**—RADSL is a variant of ADSL that overcomes varying conditions and lengths of copper cable. RADSL has the same maximum data rates as ADSL, but both downstream and upstream rates are adjusted to the line conditions at the time.

- **Very High Bit Rate DSL (VDSL)**—VDSL provides downstream rates of approximately 13 to 52 Mbps, and upstream rates of up to 16 Mbps. The maximum distance between the CO and customer is much shorter than other DSL varieties, as little as 1,000 feet for the highest rate. However, since VDSL can run over fiber optic cable, a telecommunications company can locate a nearby gateway to connect fiber from the CO and copper to the house or business. This all but eliminates distance limitation concerns. It is particularly useful for campus settings and businesses. VDSL is being introduced for market trials to deliver video services over existing phone lines. VDSL can also be configured in symmetric mode.

The equal transmission speeds of Symmetrical DSLs are useful for LAN access, video-conferencing, and for entities hosting their own Web sites. Symmetrical DSL technologies include:

- **Symmetric or Single Line DSL (SDSL)**—SDSL offers symmetric upstream and downstream speeds of up to 2 Mbps or greater over one wire pair. Business-grade SDSL eliminates the POTs voice component found in ADSL in order to use the entire line bandwidth for data. SDSL is more appealing as an alternative to T1 service for businesses that need generous upstream and downstream bandwidth (for Web hosting, WAN connectivity, and so forth). SDSL is much more expensive than the equivalent ADSL service (384/384 Kbps SDSL costs approximately $130/month, where 1.5 Mbps down/384 Kbps upstream ADSL costs approximately $70/month), and so is not a good alternative for home use.

- **High Data Rate DSL (HDSL)**—HDSL was created in the late 1980s to deliver service at speeds up to 2.3 Mbps in both directions. HDSL uses as many as three twisted copper pairs.

- **SHDSL (Symmetric High-Bitrate DSL)**—SHDSL is state-of-the-art, industry standard symmetric DSL. SHDSL provides 20 percent better loop-reach than older versions of symmetric DSL, causes much less crosstalk, and provides multi-vendor interoperability. SHDSL runs at bit-rates from 192 Kbps to 2.3 Mbps. SHDSL can use one or two pairs of copper wires. SHDSL is being adapted for voice-over-DSL.

- **2nd Generation HDSL (HDSL2)**—HDSL2 delivers 1.5 Mbps transmission service each way supporting voice, data, and video, using either asynchronous transfer mode (ATM), private-line service, or frame relay over a single copper pair.

- **Integrated Services Digital Network DSL (IDSL)**—IDSL uses ISDN line coding to transport data-only services at 144 Kbps. IDSL is generally much more expensive than other equivalent DSL services, but may be the only option when the home or office is too far from the CO for other services. IDSL can operate through repeaters and digital loop carrier (DLC)-based circuits and eliminates the call setup steps required on dial-up ISDN circuits."

Physical Requirements

DSL comes in a variety of speeds, which makes it attractive to the consumer. However, it depends on the physical condition of the copper loop over which it is provisioned.

- **Reach (loop length)**—Each xDSL service has a maximum reach, or distance the service can be offered from the CO. Typically, the range of DSL is between 12,000 and 26,000 feet. Customers located beyond the reach of a service cannot receive the service.

- **No devices**—The copper wires between CO and customer must be free of electronic devices such as repeaters. This is a problem because most copper that extends more than 6,000 feet from the CO contains some kind of electronic repeater to boost the analog signal. It is also essential that the line does not contain loading coils. These were typically installed on analog lines to filter out high-frequency noise by cutting off all frequencies above 4 kHz. However, if we look back at the ADSL Bandwidth Diagram, we will see that such a device would block the wide range of frequencies used by xDSL.

Thus, an incumbent local exchange carrier (ILEC) must first remove these devices before it can provide DSL service to many of its customers. This increases the time and cost of many xDSL installations.

- **Good wiring**—The copper loop must also be in good physical condition, well-installed, with no mismatched wire gauges. The age of the average copper wire plant makes this unlikely. Bellcore estimates the typical U.S. telephone line crosses 22 splices, which allows line noise and crosstalk to reduce effective data rates. In addition, effective data rates are reduced by other problems throughout the telephone system, such as overlong loops that attenuate signals, nonterminated wire pairs, and crosstalk between wires.

ADSL

Asymetrical Digital Subscriber Line (ADSL) is called "asymmetric" because its downstream data rate is much faster than its upstream rate. This type of DSL is one of the most attractive solutions for home users, because it matches typical Internet usage patterns. For example, a Web surfer only sends about 10 keystrokes upstream to bring a large Web page downstream.

The majority of target applications for digital subscriber services are asymmetric. VOD, home shopping, Internet access, remote LAN access, multimedia access, and specialized PC services all feature high data rate demands downstream, to a subscriber, but relatively low data rate demands upstream. For example, Motion Picture Experts Group (MPEG) movies with simulated VCR controls require 1.5 or 3.0 Mbps downstream; however, they work just fine with no more than 64 Kbps (or 16 Kbps) upstream. IP protocols for Internet or LAN access push upstream rates higher; however, a 10 to 1 ratio of downstream to upstream rates does not compromise performance in most cases.

The reason ADSL is asymmetrical has less to do with transmission technology than with the cable plant itself. Twisted pair telephone wires are bundled together in large cables, usually with 50 wire pairs to a cable. However, cables coming out of a CO may have hundreds or even thousands of pairs bundled together.

Thus, an individual line from a CO to a subscriber is spliced together from many cable sections as they fan out from the CO. Alexander Graham Bell invented twisted pair wiring to minimize the interference of signals from one cable to another caused by radiation or capacitive coupling; however, the process is not perfect. All of this interference, bundling, and splicing limits the local loop to approximately 1.1 megahertz (MHz) of bandwidth.

ADSL has a range of downstream speeds, depending on distance between the CO and a subscriber, as presented in the ADSL Data Rates Table.

ADSL Data Rates

Data Rate (Mbps)	Wire Gauge (AWG)	Distance (feet)	Wire Size (millimeter)	Distance (kilometer)
1.5 or 2	24	18,000	0.5	5.5
1.5 or 2	26	15,000	0.4	4.6
6.1	24	12,000	0.5	3.7
6.1	26	9,000	0.4	2.7

Upstream speeds range from 16 to 640 Kbps. Individual products today incorporate a variety of speed arrangements, from a minimum set of 1.544/2.048 Mbps downstream and 16 Kbps upstream to a maximum set of 9 Mbps downstream and 640 Kbps upstream. All of these arrangements operate in a broad frequency band above the 4,000-Hz POTS voice band, leaving POTS independent and undisturbed even if a premise's ADSL modem fails. The ADSL Connectivity Diagram illustrates typical connectivity for this type of subscriber line.

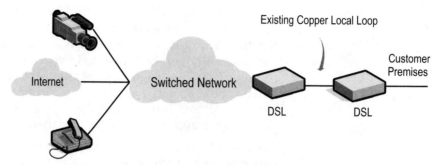

ADSL Connectivity

The ADSL Configuration Diagram provides an overview of an ADSL network. An ADSL circuit connects an ADSL modem on each end of a twisted pair telephone line. It creates three information channels as follows:

- High-speed downstream channel that connects to an ATM network

- Medium-speed duplex channel

- POTS channel, which is split off from the digital system by filters, thus guaranteeing uninterrupted POTS, even if ADSL fails

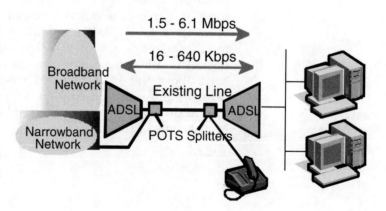

ADSL Configuration

ADSL uses analog signals, but spreads them out over a range of frequencies 100 or more times greater than dial-up modems. The spectrum is sliced into dozens of narrow bands, as if 100 modems were sending signals over one wire simultaneously.

ADSL is considered the most viable version of DSL, because it works over long distances. Downstream data rates depend on several factors: length of the copper line, its wire gauge, presence of bridged taps, and cross-coupled interference.

Because many applications planned for ADSL involve a real-time signal, link and network-level error control protocols cannot be used. Therefore, ADSL modems incorporate forward error correction (FEC).

To create transparent multiple channels at various data rates, ADSL modems divide the available bandwidth by either FDM or echo cancellation, as illustrated on the ADSL Channel Diagram.

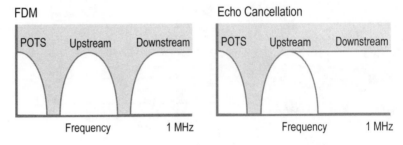

ADSL Channel

- FDM assigns one band for upstream data and another band for downstream data. TDM further divides the downstream path into one or more high-speed channels and one or more low-speed channels. The upstream path is also multiplexed.

- Echo cancellation allows two stations to transmit on the same band (simultaneously upstream and downstream). To create two separate transmission channels, each station subtracts, or cancels, its own transmission from the combined signal, so that it receives only the signal transmitted by the other station.

An ADSL modem multiplexes downstream channels, upstream channels, and maintenance channels together in blocks, attaching an error code to each block. A receiver corrects errors occurring in the transmission, up to the limits of the code and block length. Technical and practical problems must be overcome to enable widespread deployment of ADSL, including adoption of standards and readiness of the local loop.

Activity

1. List the advantages and disadvantages of the following technologies:

   a. 56-Kbps modem

   _____

   _____

   _____

   _____

   b. ISDN

   _____

   _____

   _____

   _____

   c. ADSL Lite

   _____

   _____

   _____

   _____

   d. RADSL

   _____

   _____

   _____

   _____

   e. ADSL

   _____

   _____

   _____

   _____

f.   Cable modem

_____

_____

_____

_____

Extended Activity

Visit the ADSL Forum Web site **http://www.dslforum.org** and research the latest developments in ADSL.

Lesson 6—Cable Modems

Cable modems are another technology aimed at the local loop. Cable modems offer much more than basic video services, providing Internet connectivity and other home and small business applications.

Objectives

At the end of this lesson you will be able to:

- Describe the basics of cable modem technology
- Compare the services of cable modems and ADSL

Key Point

Cable modems provide video, telephone, and data services to a subscriber.

Cable Modem Technology

The Internet via Cable Modem Diagram provides an overview of cable modem access over an HFC network. HFC networks are composed of fiber feeder from a cable head end to a neighborhood optical node serving several hundred homes. A signal then travels from the node to each home over coaxial cable. A network interface unit (NIU) inside the home includes a cable modem and other electronics, and perhaps a power supply.

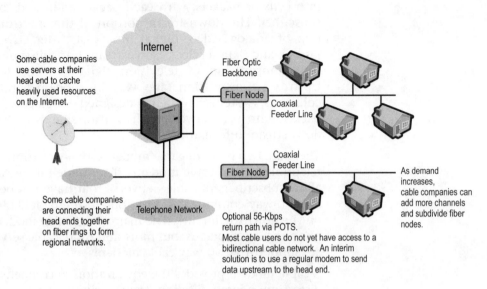

Internet via Cable Modem

A cable modem is a complex device that incorporates a tuner, which separates data signals from broadcast streams (video) from telephony; parts from network adapters, bridges, and routers; network management software agents, which enable a cable company to monitor operations; and encryption devices. Each cable modem has an Ethernet port. As a result of this configuration, up to three cable wires can be provided from the NIU: a coaxial wire delivering broadcast video to a television, an Ethernet wire connecting to a PC, and a twisted pair of wires connecting to a telephone. TCP/IP software is required in a computer.

The current HFC cable network uses a 750-MHz spectrum, equivalent to 110 downstream television channels of 6 MHz each (a Federal Communications Commission [FCC] limit). The most common cable modems create a downstream data stream out of one of the 6-MHz television channels that occupy the spectrum between 50 and 750 MHz. Using 64 quadrature amplitude modulation (QAM), downstream transmission can occur at rates as high as 30 Mbps.

Spectrum bandwidth from 5 to 42 MHz is reserved for upstream signaling and telephony. Common cable modems create an upstream channel out of this currently unused band.

The downstream channel is continuous; however, it is divided into cells or packets, with each packet addressed to a particular subscriber. The downstream portion of the spectrum supports a mix of analog video, digital broadcast, interactive video, telephone, and data services. Downstream transmission does not disturb transmission of television signals to the television set. Upstream transmission rates vary by modem vendor. To avoid collisions, systems are being designed to place each upstream packet onto the network with control signals embedded in the downstream information stream.

There are two types of cable modems: two-way cable modems and telephone return cable modems. Telephone return cable modems allow subscribers of cable networks that have not been upgraded for two-way communication to obtain the benefits of high speed on the downstream link. The majority of cable modems were telephone return modems, but more and more cable service providers are supporting two-way cable modems.

A cable modem provides the equalization to compensate for signal distortion, address filtering, transmitting and receiving functions, automatic power adjustments, adjustments in amplitude (to compensate for temperature changes), signal modulation, and compensation for delays caused by variable distances from the head end.

Cable providers must overcome several technical and practical problems to enable widespread deployment of cable modems. These include technical standards for customer premises equipment (CPE) and readiness of the cable plant for two-way traffic. In addition, on the upstream path, analog noise problems are sigificant and difficult to resolve.

ADSL or Cable Modem? The Subscriber's Perspective

From the end-user's perspective, both ADSL and cable modem offer a continuous connection, thus making the Internet as immediately accessible as a compact disc-read only memory (CD-ROM) drive. Both currently require a visit by an installer. The choice of ADSL or cable modem depends upon the user's preferences regarding shared bandwidth, price, and choice of host (Internet service provider [ISP] or corporate LAN).

- **Shared bandwidth**—ADSL involves a dedicated connection, while cable modem requires users to share bandwidth in a traditional Ethernet broadcast network. If many cable modem users are online, downstream speeds can fall as low as 64 Kbps, substantially below the advertised 10 Mbps. Cable companies intend to install more head-end equipment as higher numbers of subscribers slow down access. A more troubling aspect of the shared cable modem network is lack of security. A marginally skilled hacker could dig his way into a neighbor's computer files. Firewalls, either as software loaded on a PC or as a component of a router, can help protect local files from Internet attacks.

- **Price**—Cost is a very important consideration for subscribers. The ADSL vs. Cable Modem Table lists sample rates for comparative home broadband services.

ADSL vs. Cable Modem

Data RatesDown/ Up	ADSL	Cable Modem
256/256 Kbps	$37.95*	$57.95**
1.5Mbps/ 896Mbps	$50.95*	$94.95***

*Includes modem rental, ISP
**Includes modem rental, speeds up to 3 Mbps down, 256 Kbps up
***Includes modem rental, speeds up to 4 Mbps down, 512 Kbps up

- **Choice of host**—A telephone company's ADSL service is based on a hub-and-spoke model, in which the hub can be either a corporate LAN or ISP. Some telecommunications providers support a range of hubs, by selling high-speed connections (1.5 to 45 Mbps) to multiple ISPs. This allows individual subscribers to choose among several ISPs for Internet access. In contrast, the leading cable modem services currently link only to their proprietary content offering, which includes Internet access. A user subscribed to those services does not have a choice of ISP.

Thus far, cable modem is winning the competitive battle to bring high-speed Internet access to the home. Cable modem has the first-mover advantage, as deployments began in early 1997.

Activities

1. Discuss the advantages and disadvantages of cable modems.

   _____

   _____

   _____

   _____

2. Contrast and compare cable modems and ADSL services.

   _____

   _____

   _____

   _____

Extended Activities

1. Research the latest developments in cable modem technologies.
2. Research the latest developments in ADSL technologies.

Lesson 7—Synchronous Optical Network

Fiber optic transmission has been used for some time in public long distance networks. Links in the first generation of fiber optics were entirely proprietary in nature, including architectures, equipment, protocols, and formats for multiplexing frames.

The SONET standard now standardizes optical transmission by providing the rules for converting electrical signals to pulses of light, moving those light pulses over thin strands of fiber optic cable, and converting the pulses back to electrical signals at their destination.

Objectives

At the end of this lesson you will be able to:

- Explain the purpose for SONET development and deployment

- Describe the protocols that make up SONET architecture

- List devices used in a SONET-based network

 **Key Point**

SONET uses the term OC as the data rate descriptor.

The SONET Standard

The SONET standard defines a signal hierarchy similar to that which we saw for T-carriers, but extending to much higher bandwidths, as presented in the SONET Bandwidth Table. The basic building block is the Synchronous Transport Signal level 1 (STS-1) 51.84-Mbps signal, chosen to accommodate a DS3 signal. The hierarchy is defined up to STS-48, that is, 48 STS-1 channels for a total of 2,488.32 Mbps, capable of carrying 32,256 voice circuits. The STS designation refers to the interface for electrical signals. The optical signal standards are correspondingly designated OC-1, OC-2, etc.

SONET Bandwidth

STS and OC	STM	Rate (Mbps)	Number of DS1s	Number of DS3s
STS/OC-1		51.84	28	1
STS/OC-3	1	155.52	84	3
STS/OC-12	4	622.08	336	12
STS/OC-48	16	2488.32	1344	48
STS/OC-192	64	9953.28	5376	192

SONET began as a U.S. standard, and was later incorporated into the international standard Synchronous Digital Hierarchy (SDH). SDH is being developed by ITU-T and many international post, telephone, and telegraph (PTT) companies. SONET/SDH is the standard that meets these worldwide needs for standardization. The difference between the two standards is shown in the table. SONET uses the term OC or STS as the data rate descriptor. SDH uses Synchronous Transport Mode (STM).

Standardization of OC services has obvious advantages to telephone companies, making it possible for them to select equipment from multiple vendors and interface with other telephone companies "in the glass," that is, without converting a signal back to copper transmission. SONET allows synchronous signals as low as DS0 to be switched without being demultiplexed.

SONET Advantages

SONET has important advantages for those who use switched networks for communications:

- The SONET standards provide a low-level platform upon which other standards can be based.

- SONET makes it possible for subscribers to purchase equipment that interfaces "in the glass" with switched public networks. For example, SONET interfaces are available to Switched Multimegabit Data Service (SMDS) and ISDN.

- Even the earlier offerings, OC-1 to OC-3, make new applications combining data, voice, and video images both technically and economically feasible.

- SONET provides direct, transparent interfaces to Layer 2 WAN protocols such as ISDN. For example, SONET appears to ISDN interfaces as a continuation of the copper-cable based ISDN network.

The SONET standard includes extensive network operation and management facilities. A significant portion of the SONET bandwidth is allocated for out-of-band control signaling for this purpose. This management system has its own OSI-compliant communications architecture. Ultimately, subscribers will be able to interface directly to this capability, running the SONET "stack" on their own computers.

SONET Protocol Architecture

The SONET and OSI Model Diagram shows how SONET architecture relates to the OSI model. Keep in mind that SONET is a Physical Layer standard; it deals with the transmission of bits of data and the electrical and optical forms of these bits. The SONET Physical Layer is divided into four layers:

- Path
- Line
- Section
- Photonic

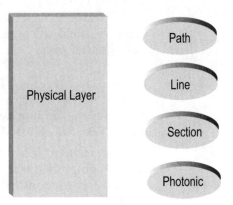

SONET and OSI Model

Path Layer   The path layer is the logical connection for end-to-end management and delivery of data. The path layer is responsible for establishing a complete path over the SONET network between end network interface devices, called service adapters. The service adapters map DS3, ISDN, Fiber Distributed Data Interface (FDDI), or other protocols onto the SONET network.

The SONET path layer is similar in concept to a Network Layer protocol. The path layer's job is to establish an end-to-end path between internetworked nodes.

The path layer is the entry and exit point for services in SONET. Through a portion of reserved bandwidth called the path overhead (POH), the path layer has responsibility for the following functions occurring between network elements:

- Map and transport of services

- Equipment status

- Connectivity

- Error monitoring

- User-defined functions

Path terminating equipment (PTE) consists of network elements that originate and terminate transported services. PTE, such as SONET digital cross-connect (DCS) systems, read, interpret, and modify the POH.

Services positioned, or mapped, into the STS-1 path layer are called the payload. The POH resides within the payload. The payload and POH can be placed anywhere within the portion of the SONET frame that carries the mapped services. This portion of the frame is called the synchronous payload envelope (SPE), which we discuss in more detail later.

Line Layer

The line layer is responsible for reliable transport of the path layer payload and POH across the transmission medium (usually fiber optic cable). The line layer multiplexes and synchronizes STS-1s into STS-3s, STS-3s into STS-9s, and so on.

Network elements operating at the line layer are called line terminating equipment (LTE). The line layer provides the following LTE-to-LTE functions for the path layer payload and POH:

- Synchronization

- Payload location

- Multiplexing

- Error monitoring

- Automatic protection switching (APS)

These functions are performed with a portion of bandwidth in the STS called the line overhead (LOH), which is read, interpreted, and modified by any equipment that terminates this layer. Examples of LTE are SONET fiber optic MUXs, including add/drop MUXs. PTE also performs the functions of LTE, just as a router functions at both the Network and Data Link Layers.

Section Layer

The section layer provides functionality similar to the Data Link Layer of the OSI model. The section layer is responsible for transporting the STS-N across fiber optic cable, building STS-N frames and moving them from the source to the destination.

Network elements at this layer are called section terminating equipment (STE). Using a reserved portion of the STS-1 called the section overhead (SOH), the section layer performs the following STE-to-STE functions:

- STS identification
- Framing
- Error monitoring
- User-defined functions

SOH is read, interpreted, and modified by all equipment that terminates this layer. Just as a router must support the lower layers, PTE and LTE perform STE functions. SONET regenerators (repeaters), used to extend SONET optical transmission distances, are examples of STE network elements.

Photonic Layer

The photonic layer is responsible for transmission of bits across optical fiber. This layer converts electrical input signals to optical signals and vice versa. A transceiver is the device used to convert electrical signals to optical signals. Optical equipment communicates at this layer, and there is no overhead.

The primary function of this layer is to convert electrical STS-N frames to light pulses for transmission as OC-N across optical fiber. Activities monitored at this level include optical pulse shaping, power levels, and wavelength.

SONET Multiplexing

SONET uses the STS-1 bit rate (51.84 Mbps) as the basic building block. Higher transmission speeds are multiples of the STS-1 rate. The SONET STS-3 Multiplexing Diagram demonstrates the basic multiplexing structure of SONET. Any type of service ranging from DS0 to high data rate switched services such as Broadband ISDN (B-ISDN) can be accepted at the SONET path layer by the service adapters. An adapter maps the signal into the payload envelope of the STS-1. New services and signals can be transported by adding new service adapters at the edge of the SONET network. In this example, three STS-1s are multiplexed, or "concatenated," into an STS-3 at the line layer, and converted into an OC-3 signal at the section layer.

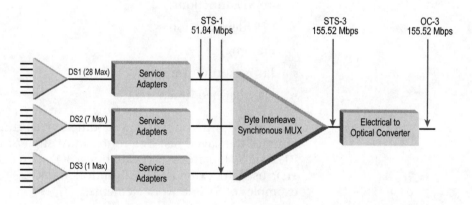

SONET STS-3 Multiplexing

Each input is eventually converted to a base format of a synchronous STS-1 signal (51.84 Mbps) or higher. Lower-speed inputs, such as DS1s, are first bit or byte multiplexed into virtual tributaries, which group these lower speed circuits together to form STS-1s. Then, as illustrated on the SONET Synchronous Multiplexing Diagram, these synchronous STS-1s are multiplexed together in either a single- or two-stage process to form an electrical STS-N signal.

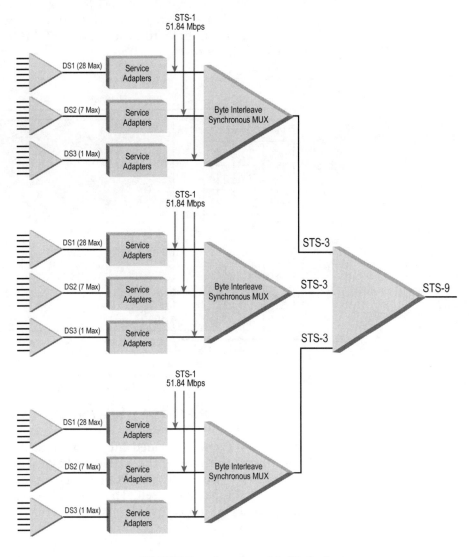

SONET Synchronous Multiplexing

STS multiplexing is performed at the byte interleave synchronous MUX. Bytes are interleaved together in a format such that low-speed signals are visible. Then a conversion takes place to convert the electrical signal to optical form, the OC-N signal.

SONET Frame Format

Each STS level has its own corresponding frame format. A SONET STS-1 frame, shown on the SONET STS-1 Frame Format Diagram, contains 810 bytes, arranged 90 bytes (columns) wide and 9 bytes (rows) deep. In other words, each row contains 90 bytes, and the frame contains 9 rows.

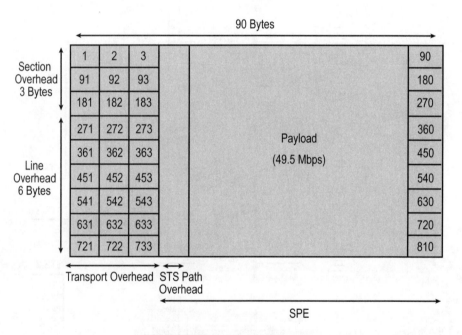

SONET STS-1 Frame Format

The SONET frame is divided into two parts: the transport overhead (first 9 columns of the frame) and SPE. The SPE can also be divided into two parts: STS POH (first column of the payload field) and the payload itself. The payload is the user data being transported and routed over the SONET network.

After the payload is multiplexed into the payload envelope, it can be transported and switched through SONET without the need for interpretation at intermediate nodes. Thus, SONET is said to be service-independent or transparent. The STS-1 payload has the capacity to transport up to:

- 28 DS1s

- 14 DS1Cs

- 7 DS2s

- 1 DS3

- 21 CEPT1s (E1-type signal)

A frame is transmitted byte-by-byte beginning with byte 1, going from left to right until byte 810 is transmitted. The entire frame is transmitted in 125 microseconds.

Each of the sections in the frame corresponds to specific "headers" in the SONET frame. The line overhead and section overhead are typically combined and represented as the transport overhead. The POH is carried as a portion of the payload. The Sample SONET Configuration Diagram shows the section, line, and path portions of a SONET network. Overhead is added for each of these portions to allow for simpler multiplexing and lower circuit maintenance. Path level overhead is carried from end to end.

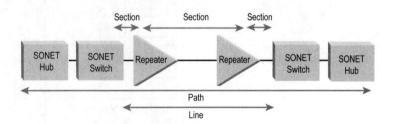

Sample SONET Configuration

Virtual Tributaries

SONET also defines sub-STS-1 levels called virtual tributaries (VTs). VTs map sub-STS-1 services, such as DS1 and DS2, into STS-1 frames. There are four VT speeds defined, as listed in the Virtual Tributaries Table. Tributaries can be viewed as inputs to a SONET-based system.

Virtual Tributaries

Type	Transports	VT Rate (Mbps)
VT1.5	1 DS1	1.728
VT2	1 CEPT1	2.304
VT3	1 DS1C	3.456
VT6	1 DS2	6.912

Within an STS-1 frame, each VT occupies a number column, as illustrated on the STS-1 Framing Diagram. Within the STS-1, many VT groups can be mixed together to form an STS-1 payload.

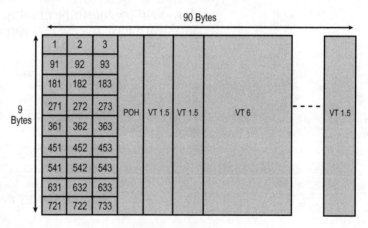

STS-1 Framing

Clock Synchronization

Clocking is a key aspect of digital communications. Clocking ensures that the sender and receiver recognize when bits appear in the data stream. SONET VTs also require clock signals for data stream synchronization. Synchronous and asynchronous multiplexing techniques are two methods in which DS1 signals can be combined into higher bit rate data streams; SONET uses synchronous multiplexing. To better appreciate SONET's synchronous multiplexing system, first consider asynchronous multiplexing.

Asynchronous multiplexing combines DS1s together into DS2s and then into DS3s. Because clock references vary between individual circuits, multiplexing techniques must allow for these variations. Asynchronous multiplexing accomplishes this by using a method called "bit stuffing," which adds bits to the data stream to fill the "gaps" between bit rates at different digital signal levels.

To access individual DSX signals, a receiver must first demultiplex the combined signals, removing the added bits as well. SONET's synchronous multiplexing techniques combine DSX VTs into STS-1 SPEs. SONET uses a set clock reference of 1.7288 Mbps per VT1.5 (DS1). This set reference value allows the STS-1 to carry each VT within the SPE so that each VT is visible to each PTE without the need for the receiver to demultiplex the STS-1 payload. The result is faster access to the DSX signals than asynchronous multiplexing techniques. An individual VT containing a DS1 can be extracted without demultiplexing the entire STS-1.

SONET Network Elements

There are several components that may be used in a SONET-based network. Some of the more common components are:

- Add/drop MUX/demultiplexer
- Broadband digital cross-connect
- Wideband digital cross-connect
- Terminating MUX
- Regenerator (repeater)

Add/Drop MUX/ Demultiplexer

An add/drop MUX/demultiplexer can multiplex various inputs into an OC-N signal. It can be used at terminal sites or intermediate network locations to insert (add) or remove (drop) DSx or OC-N signals into an existing OC-N signal. It is configured as a hub, as shown on the Add/Drop MUX Diagram. At an add/drop site, only those signals that need to be accessed are dropped or inserted. The remaining traffic continues straight through without interruption, and does not require special equipment or extra processing. An add/drop MUX is considered LTE.

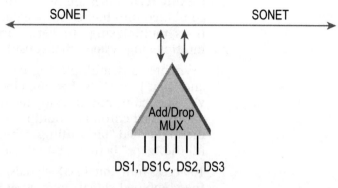

SONET   SONET

DS1, DS1C, DS2, DS3

Add/Drop MUX

Broadband Digital Cross-Connect

A broadband digital cross-connect switches STS-1s within various OC rates. The SONET Broadband Digital Cross-Connect Diagram depicts this component. One major difference between a cross-connect and add/drop MUX is that a cross-connect provides the capacity to interconnect a much larger number of STS-1s than an add/drop MUX.

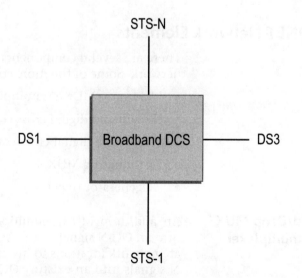

STS-N

DS1 —— Broadband DCS —— DS3

STS-1

SONET Broadband Digital Cross-Connect

Wideband Digital Cross-Connect

A wideband digital cross-connect is similar to the broadband digital cross-connect, except that it switches at VT levels, as illustrated on the SONET Wideband Digital Cross-Connect Diagram.

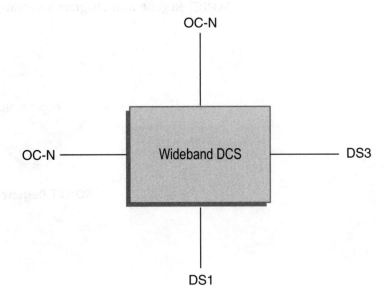

SONET Wideband Digital Cross-Connect

The wideband digital cross-connect accepts DS3s and DS1s. Only the required VTs are accessed and switched, leaving the OC-N signals intact, allowing for more granular multiplexing/demultiplexing than that allowed by a broadband digital cross-connect. The wideband digital cross-connect is considered PTE.

Terminating MUX

Terminating MUXs are devices used to access a SONET network, as illustrated on the SONET Terminating MUX Diagram. The terminating MUX (TMUX) is considered PTE, and is used as an entry-level access point to the edge of the SONET network.

SONET Terminating MUX

Regenerator

A SONET regenerator increases the optical signal level of the OC-N when long distances between devices dictate its use. The regenerator replaces the SOH in the received STS-N frames, but leaves the LOH, POH, and payload intact. Thus, a regenerator is STE. The SONET Regenerator Diagram illustrates this concept.

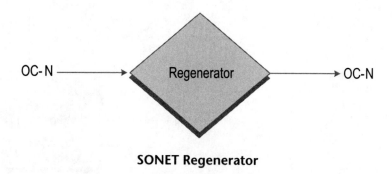

SONET Regenerator

Activities

1. Why was 51.84 Mbps chosen as the fundamental rate for SONET?

    _____

    _____

2. What is the difference between SDH and OC?

    _____

    _____

3. Considering the SONET STS-N data rates, why would ATM normally operate at 155 and 622 Mbps?

    _____

    _____

4. At which OSI layer would you classify SONET?

    _____

    _____

Extended Activities

1. What technologies can SONET support? Is it possible to carry voice conversations over SONET? How would a voice network interface with a SONET network? How would you interface an IP network to SONET? What devices would you use?

2. Research which Physical Layer protocols are used for the Internet backbone.

Summary

Module 3 began by reviewing the concepts of physical and logical circuits. While a physical circuit is an unchanging, tangible connection between two points, a logical circuit is essentially a set of instructions that moves data between two points by means of one or more intermediate nodes. Physical circuits are configured from the point-to-point technologies covered in this module. Logical circuits, such as SVCs and PVCs, are set up by means of switched services, which will be discussed in Module 4.

The smallest and slowest of the point-to-point offerings is the dial-up analog line, with a modem at each end. This is essentially the same type of service available to individual home users, although data rates are available up to 56 Kbps. If a home or business needs data transfer rates up to 56 Kbps, DDS offers an affordable, all-digital solution for small organizations. However, it is only available in areas fairly near a CO.

After a user has enjoyed the benefits of all-digital transmission, the large family of T-carriers provides a series of step-up services. Customers who only need a few digital lines can purchase individual channels of a T1 line. Each 64-Kbps channel, or FT1, can transmit either voice or data. Businesses commonly use some T1 channels for PBX connections (tie lines), and others to link LANs over a wide or metropolitan area. The main disadvantage of FT1 is the inability to choose its endpoints. Because other customers are sharing the physical transmission path, each FT1 links a customer location to the nearest telephone company CO.

A full T1 provides 24 channels of 64 Kbps each (1.544 Mbps total), plus the flexibility of choosing both endpoints of the connection. T1 channels can also be bonded, or recombined into a smaller number of higher-capacity channels. A T3 line multiplexes 28 T1 lines, for a total of 672 channels.

While T-carriers offer cost-effective connectivity for large organizations and businesses, most small businesses and home computer users still connect to the Internet using modems over analog local loops. DSL technology offers high-speed switched digital transmission over old copper wires. One of the most promising DSL "flavors" is ADSL, which meets the needs of Internet users by providing a very large downstream channel for fast Web page access and file downloads. However, all types of DSL have strict physical requirements, such as distance from a CO, that limit the number of subscribers who can take advantage of them.

Some high-speed home services solve this problem by ignoring the telephone loop altogether. Cable television companies now offer Internet access over the same coaxial networks used for private television transmission. Although cable modem technology can potentially provide fast data rates at a reasonable price, these networks use shared-medium access methods similar to Ethernet. Thus, actual rates depend on the number of simultaneous users, and the network is vulnerable to common security problems.

At the top of the point-to-point hierarchy, the SONET standard provides high-speed digital transmission over optical fiber. While many digital protocols are designed for wide area use only, the SONET protocols extend all the way down to the desktop level. In theory, an organization could use SONET for both local and wide area communication. In practice, SONET provides the basic transmission service for wide area switched networks such as ATM.

Module 3 Quiz

1. An Ethernet LAN needs to connect to the Internet. The company has 20 users and has chosen DSL for their connectivity. Which piece of equipment is needed for their Internet connection?

   a. Router

   b. Digital modem

   c. DSU/CSU

   d. Hub

2. A small communications company wants to connect to the Internet and they currently have a 28.8-Kbps modem. They are not very close to a CO. They need to upgrade their speed of connectivity, but cost is a major issue. What would be their best solution?

   a. 56-Kbps leased line

   b. xDSL

   c. T1

   d. T3

3. A company wants to connect to the Internet. The primary use of the Internet connection will be e-mail and Web research. Scalability of the Internet speed is of primary importance to the customer. What type of connection will allow for this scalability?

   a. ISDN-BRI

   b. ISDN-PRI

   c. 56-Kbps leased line

   d. xDSL

4. ADSL is best characterized by:

   a. Analog-to-digital conversion at the local loop

   b. Digital-to-analog conversion at the local loop

   c. High speed to the subscriber, low speed from the subscriber

   d. High data transfer from the subscriber, low-speed transfer to the subscriber

5. Which of the following types of telecommunications circuits are typically fixed establishments over a fixed path?

   a. Switched line

   b. Dial-up line

   c. Full-duplex line

   d. Leased line

6. An example of a point-to-point alternative used to transmit data over a wide area might be:

   a. Leased lines

   b. X.25

   c. ISDN

   d. ATM

7. FT1 is a multiple of:

   a. 64-Kbps channels

   b. 58-Kbps channels

   c. T1 channels

   d. T3 channels

8. DDS uses which of the following?

   a. Satellite communications

   b. Digital modems

   c. Analog modems

   d. Codecs

9. T1 is equivalent to:

   a. DS0

   b. ISDN Basic Rate

   c. DS1

   d. E1

10. The building block for SONET is:

    a. STS-1

    b. 51.84 Mbps

    c. 48 Kbps

    d. 64 Kbps

    e. Both a and b

11. When ATM is carried at 622-Mbps rates, the physical transport would be:

    a. DDS

    b. T1

    c. T3

    d. SONET

12. SONET is found at which of the following layers?

    a. Physical

    b. Data Link

    c. Network

    d. Transport

Module 4
Switched Telecommunications Protocols

In contrast to the dedicated digital services described in Module 3, this module reviews protocols closely associated with the Data Link Layer of the Open Systems Interconnection (OSI) model. Some protocols also provide features associated with the Network Layer. In other words, some of these protocols, such as Integrated Services Digital Network (ISDN), are used to move information across one physical link. Other protocols, such as frame relay and Asynchronous Transfer Mode (ATM), move information across multiple links in a network.

In this module, we examine protocols commonly used to create "any-to-any" wide area connectivity without the expense of connecting each pair of locations with a dedicated private line.

Lessons

1. Circuit- and Packet-Switched Networks
2. Serial Line Internet Protocol and Point-to-Point Protocol
3. Integrated Services Digital Network
4. Asynchronous Transfer Mode
5. Frame Relay
6. Switched Multimegabit Data Service
7. X.25

Terms

Asynchronous Transfer Mode (ATM)—ATM is a connection-oriented cell relay technology based on small (53-byte) cells. An ATM network consists of ATM switches that form multiple virtual circuits to carry groups of cells from source to destination. ATM can provide high-speed transport services for audio, data, and video.

Burst Range—Burst range refers to network traffic outside the range of the CIR.

Bursty—A network traffic pattern in which a lot of data is transmitted in short bursts at random intervals is referred to as bursty.

Class of Service (CoS)—CoS provides preferential treatment to certain data types on the network. IEEE 802.1p allows administrators to define up to eight separate traffic CoSs, each controlling the type of service each frame experiences on the network.

Cloud—Any switched network that provides service while hiding its functional details from its users is referred to as a cloud. A user simply connects to the edge of the cloud, and trusts the network to handle the details of moving a signal or data across to its destination. The public-switched telephone system and Internet are two well-known examples of cloud networks.

Committed Information Rate (CIR)—CIR is the guaranteed average data rate for a frame relay service.

Common Channel Signaling (CCS)—CCS dedicates a separate communications channel to control signaling, which eliminates problems associated with control signaling within a voice channel. CCS has culminated with ITU-T SS7, also called SS #7, which will gradually be adopted by most networks.

Constant Bit Rate (CBR)—CBR information requires synchronization between sender and receiver and a specified bandwidth to make sure information is communicated accurately. CBR service is used by voice, video, and similar time-sensitive traffic. The term CBR is typically associated with protocols designed for handling multimedia traffic, such as ATM.

Cyclic Redundancy Check (CRC)—CRC is the mathematical process used to check the accuracy of data being transmitted across a network. Before transmitting a block of data, the sending station performs a calculation on the data block and appends the resulting value to the end of the block. The receiving station takes the data and CRC value, and performs the same calculation to check the accuracy of the data.

Data Service Unit/Channel Service Unit (DSU/CSU)—A DSU/ CSU is the hardware required to connect a common carrier connection (leased line) to a router. A DSU takes information from a LAN device and creates digital information suitable for public transmission facilities. A CSU is the device that actually generates the transmission signals on the local loop (telephone channel). DSUs are normally coupled with CSUs in one device called a DSU/ CSU.

Dumb Terminal—A terminal that totally depends on a host computer for processing capabilities is referred to as a "dumb" terminal. Dumb terminals typically do not have a processor, hard drive, or floppy drives; only a keyboard, monitor, and method of communicating to a host (usually through some type of controller).

E1—E standards are the European standards similar to the North American T-carrier standards. E1 is similar to T1.

Frame Relay—Frame relay is a packet-forwarding WAN protocol that normally operates at speeds of 56 Kbps to 1.5 Mbps.

High-Level Data Link Control (HDLC)—The HDLC protocol suite represents a wide variety of link layer protocols such as SDLC, LAPB, and LAPD.

Interexchange Carrier (IXC)—An IXC is a long distance company (such as AT&T or MCI) that provides telephone and data services between LATAs.

International Telecommunication Union-Telecommunications Standardization Sector (ITU-T)—ITU-T is an intergovernmental organization that develops and adopts international telecommunications standards and treaties. ITU was founded in 1865 and became a United Nations agency in 1947.

Link Access Procedure Balanced (LAPB)—LAPB (or LAP-B) is an HDLC protocol subset used primarily in X.25 communications.

Link Access Procedure for D Channel (LAPD)—LAPD (or LAP-D) is part of the ISDN layered protocol. It is very similar to LAPB. LAPD defines the protocol used on the D channel to interface with a telephone company's SS7 network for setting up calls and other signaling functions.

Local Access and Transport Area (LATA)—LATAs are geographic calling areas within which a LEC may provide local and long distance services. LATA boundaries, for the most part, fall within states and do not cross state lines. Each LATA is identified by a unique area code.

Local Exchange Carrier (LEC)—A LEC is a company that makes telephone connections to subscribers' homes and businesses, provides telephone services, and collects fees for those services. The terms LEC, ILEC, and RBOC are equivalent.

Modified Final Judgement (MFJ)—In the context of this course, MFJ is the court decision, effective January 1, 1984, that required AT&T to split apart and divest itself of its 22 local telephone companies.

Next Hop Resolution Protocol (NHRP)—NHRP is used by routers to dynamically discover the MAC address of other routers and hosts connected to an NBMA network. These systems then can communicate directly without requiring traffic to use an intermediate hop, increasing performance in ATM, frame relay, SMDS, and X.25 environments.

Non-Broadcast Multi-Access (NBMA)—NBMA describes a multi-access network that either does not support broadcasting (X.25, for example) or in which broadcasting is not feasible, such as an SMDS broadcast group or an extended Ethernet that is too large.

Non-Facility Associated Signaling (NFAS)—NFAS is a form of ISDN-PRI out-of-band signaling that allows a single D-channel to control multiple PRIs. This allows all but one of the physical T1 carriers to support 24 ISDN B-channels, requiring that only one PRI dedicate a signaling channel.

Permanent Virtual Circuit (PVC)—A PVC is a connection across a frame relay network, or cell-switching network, such as ATM. A PVC behaves like a dedicated line between source and destination end-points. When activated, a PVC will always establish a path between these two end points.

Private Signaling System Number 1 (PSS1 or Q.SIG)—PSS1 is an ISO standard that defines the ISDN signaling and control methods used to link PBXs in private ISDN networks. The standard extends the "Q" point in the ISDN logical reference model, which was established by the ITU-T in its Q.93x series of recommendations that defined the basic functions of ISDN switching systems. Q.SIG signaling allows certain ISDN features to work in a single- or multivendor network.

Q.SIG—See PSS1.

Request for Comment (RFC)—An RFC is one of the working documents of the Internet research and development community. A document in this series may be on essentially any topic related to computer communication, from a meeting report to the specification of a standard.

Signaling System 7 (SS7)—SS7 (also called SS #7) is an out-of-band system that exchanges control signals and call routing information between CO switches. It is a separate network that connects all COs, regardless of where they are or who they belong to.

Statistical Multiplexing—Statistical multiplexing, or STDM, is a more flexible method of TDM. TDM allocates a fixed number of time slots to each channel, regardless of whether the channel has data to send. In contrast, a statistical multiplexer analyzes transmission patterns to predict "gaps" in a channel's traffic that can be temporarily filled with part of the traffic from another channel.

Switched Multimegabit Data Service (SMDS)—SMDS is a connectionless service used to connect LANs, MANs, and WANs at rates up to 45 Mbps. SMDS is cell-oriented and uses the same format as the ITU-T B-ISDN standards. The internal SMDS protocols are called SIP-1, SIP-2, and SIP-3. They are a subset of the IEEE 802.6 standard for MANs, also known as DQDB.

T1, T3—T1 and T3 are two services of a hierarchical system for multiplexing digitized voice signals. The first T-carrier was installed in 1962 by the Bell System. The T-carrier family of systems now includes T1, T1C, T1D, T2, T3, and T4 (and their European counterparts E1, E2, etc.). T1 and its successors were designed to multiplex voice communications. Therefore, T1 was designed such that each channel carries a digitized representation of an analog signal that has a bandwidth of 4,000 Hz. It turns out that 64 Kbps is required to digitize a 4,000-Hz voice signal. Current digitization technology has reduced that requirement to 32 Kbps or less; however, a T-carrier channel is still 64 Kbps. A T1 line offers bandwidth of 1.544 Mbps; a T3 offers 44.736 Mbps.

Time-Division Multiplexing (TDM)—TDM is a multiplexing technology that transmits multiple signals over the same transmission link, by guaranteeing each signal a fixed time slot to use the transmission medium.

Type of Service (ToS), Quality of Service (QoS)—Users of the Transport Layer specify QoS or ToS parameters as part of a request for a communications channel. QoS parameters define different levels of service based on the requirements of an application. For example, an interactive application that needs good response time would specify high QoS values for connection establishment delay, throughput, transit delay, and connection priority. However, a file transfer application needs reliable, error-free data transfer more than it needs a prompt connection, thus it would request high QoS parameters for residual error rate/probability.

Virtual Circuit (VC)—A VC is a communication path that appears to be a single circuit to the sending and receiving devices, even though the data may take varying routes between the source and destination nodes.

Virtual Private Network (VPN)—A connection over a shared network that behaves like a dedicated link is referred to as a VPN. VPNs are created using a technique called "tunneling," which transmits data packets across a public network, such as the Internet or other commercially available network, in a private "tunnel" that simulates a point-to-point connection. The tunnels of a VPN can be encrypted for additional security.

Lesson 1—Circuit- and Packet-Switched Networks

Thus far in this course, we have seen different concepts and components used to access a wide area network (WAN). We have also looked at Physical Layer protocols, which move bits across a physical media. The main concern at the Physical Layer is moving analog and digital information across a cable or through the air.

Now we move up the protocol stack to the Data Link Layer, where the sequence of Physical Layer bits provides meaning to the devices on each end of a link.

Objectives

At the end of this lesson you will be able to:

• Describe basic High-Level Data Link Control (HDLC) communication

• Describe the differences between circuit- and packet-switched networks

Key Point

Data Link Layer headers are used to move WAN frames across a link.

Data Link Layer Protocols

The protocols at the OSI Data Link Layer are the ones used to get information "into the box." In other words, Data Link protocols deliver data to a particular physical location, a computer or other network node, so that it can in turn be delivered to processes at the higher layers.

In that regard, these protocols are normally less complex than those above them. Whereas we may find many fields in a Transmission Control Protocol (TCP) message because of the various services offered by the Transport Layer, the function of Data Link Layer protocols is essentially limited to local, group, and broadcast address recognition. The Data Link Layer may also provide basic error handling and recovery mechanisms, such as frame retransmission. In addition, there are both connection-oriented and conectionless modes of operation available.

Reliable WAN Networks

WAN technologies and their corresponding protocols are used where geographically dispersed networks and subnetworks need to be logically and physically interconnected. Originally, WAN network speeds were low compared to local area network (LAN) speeds; however, improving carrier technologies are raising transfer rates substantially. Whereas 9,600- and 19,200-baud WANs were commonplace in earlier days, Physical Layer services such as T1 (1.544 megabits per second [Mbps]) and T3 (45 Mbps) are predominantly used in today's WANs.

In addition, newer, more efficient protocols are being used in WANs. For example, the Layer 2 frame relay protocol takes advantage of the extreme reliability of the newer carrier technologies. Frame relay is often found riding on T1 (or E1, the European standard) carriers, and is often used as the protocol between network routers, such as 3Com Corporation, Cisco Systems, and Nortel Networks.

Wide area networking is generally grouped into two categories:

- **Packet-switched**—Information packets, carrying full addressing and data, are sent over paths established between end nodes. The path each packet takes may vary as the packet travels to the destination device.

- **Circuit-switched**—After a circuit is established, similar to a telephone call, no further connection protocols are necessary. The session then takes place with very little addressing overhead. A certain amount of addressing is still required because multiple stations can exist in a multipoint network.

Packet-Switched Networks

Packet-switched networks are also called connectionless, because no connection is established. As in Internet Protocol (IP) networks, packets of information are passed from node to node, with a packet possibly traversing many nodes before it arrives at its destination. Many packets can be moving between the same nodes simultaneously. The Packet-Switched Network Diagram illustrates the type of network configuration.

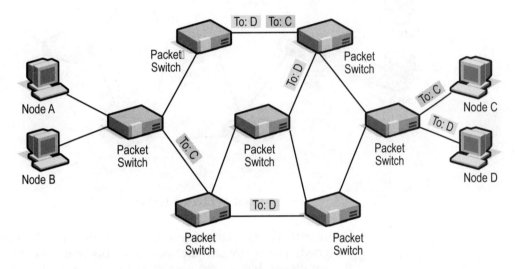

Packet-Switched Network

Note that packet-switched networks still need Data Link Layer and Physical Layer protocols to move information over each link between devices. These underlying protocols may be connectionless or connection-oriented.

Circuit-Switched Networks

Circuit-switched networks establish a single physical path between two nodes. Data packets are passed between nodes by "switching" them across this path, through intermediate nodes when necessary. Circuit-switched networks are analogous to the voice telephone system, and such connections are often called virtual circuits. A virtual circuit establishes a single route for data, and this route does not vary for the life of the connection. Therefore, circuit-switched networks are connection-oriented.

The public telephone system, frame relay, and ATM are all examples of circuit-switched networks. The Circuit-Switched Network Diagram illustrates this type of network configuration.

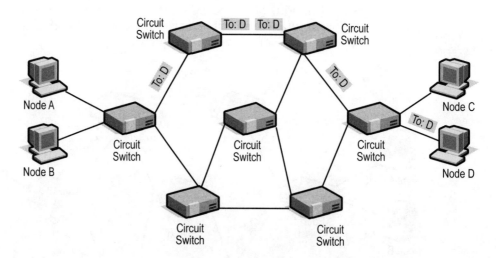

Circuit-Switched Network

In the early days of data communications, all networks were circuit switched, and many still are. For networks that cover wide areas, the emphasis has recently shifted to packet switching, simply because it permits interconnection of many more nodes into a single network. With packet switching, fewer communication channels are required (because channels are shared by many users), and interconnection of networks is much easier to accomplish. The Internet is a good example of a large packet-switching network.

Layering of data communications protocols will be discussed later in this module. As we will see, layering results in situations where the lower layers of a network are connectionless, but the higher layers establish a connection. For example, TCP is a connection-oriented protocol that uses the services of connectionless IP. The opposite scenario, where the lower layers establish a connection and the upper layers are connectionless, can also occur.

Activities

1. The protocols at the Data Link Layer _____.
   a. Physically transport data across the network
   b. Deliver data to a particular physical location
   c. Address and route data across the network
   d. Manage the flow control of data across the network

2. Packet-switched networks _____.
   a. Are connectionless
   b. Are connection-oriented
   c. Can be either connectionless or connection-oriented
   d. Transport data at the Network Layer

3. The Internet is an example of _____.
   a. A packet-switched network]
   b. A Network Layer network
   c. A circuit-switched network
   d. All of the above

Extended Activities

1. Find out how your organization connects to the Internet, and what Physical Layer and Data Link Layer protocols are used on the LAN and WAN sides.

2. Determine if these protocols are connectionless or connection oriented.

3. Go to the Web sites of the vendors mentioned in this lesson and see what Physical and Data Link Layer protocols each support in their router products.

Lesson 2—Serial Line Internet Protocol and Point-to-Point Protocol

Serial Line Internet Protocol (SLIP) and Point-to-Point Protocol (PPP) are used to transfer IP packets across a serial link. These two protocols are widely used by Internet service providers (ISPs) to provide dial-up Internet connections for home or business users.

Objectives

At the end of this lesson you will be able to:

- Describe the difference between SLIP and PPP
- Understand the basic concepts of SLIP and PPP

 **Key Point**

SLIP and PPP are used to move IP packets across a serial link.

SLIP

SLIP dates back to the early 1980s, when it was originally implemented in Berkeley Software Distribution (BSD) 4.2 UNIX. It is a simple encapsulation of an IP datagram asynchronously transmitted over serial lines using an RS-232 interface.

A typical SLIP connection is shown on the SLIP Access Diagram. A home or small business user typically uses a modem to access Internet services by means of an ISP. At the user's computer, the IP packet is placed in a SLIP (or PPP, described later) frame and sent to the modem. The modem transmits the information across the telephone network to the ISP modem. The ISP's modem is attached to a router that decapsulates the SLIP frame. The router then encapsulates the original IP packet in a WAN protocol frame, and routes it across the Internet to the proper destination.

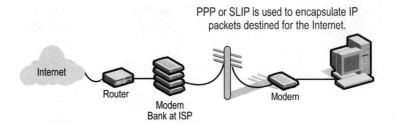

SLIP Access

SLIP Frame Delimiters

The same control character (hexadecimal C0) delimits the beginning and end of each SLIP frame. The SLIP Diagram illustrates how the IP packet (datagram) is inserted into the SLIP frame, bracketed by two hexadecimal "C0" characters.

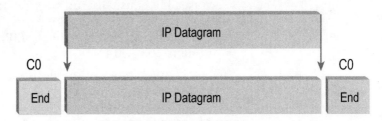

SLIP

What happens if the actual data contains "C0"? In that case, the escape (ESC) character (hexadecimal DB) is used to indicate where the data itself contains a "C0," so that the receiver will not interpret those occurrences as frame delimiters. If a "C0" appears in the data, it is transmitted as a two-character sequence of "DB" "DC." If the SLIP escape character itself appears in the data, it is represented by a "DB" "DD" sequence.

Drawbacks to SLIP

There are several drawbacks to SLIP:

- Each end must know the other end's IP address, because there is no way to exchange this information in the protocol.

- There is no "type" field that could be used to direct the data to one of several protocol stacks, thus a SLIP connection can only support one network protocol at a time.

- There is no checksum to allow for error detection on noisy telephone lines, which means the higher layers are responsible for error detection and recovery.

Despite these limitations, SLIP is a proven protocol that is very easy to implement. Its simplicity is very attractive on slow links, such as analog local loops.

Compressed SLIP

Because SLIP is usually run over relatively slow-speed serial lines, and often used for applications such as Telnet, a compressed version called compressed SLIP (CSLIP) was specified (Request for Comment [RFC] 1144).

Telnet is an interactive application, and can be very inefficient when sending just a few bytes at a time. A typical TCP/IP connection sends source and destination addresses, and various indicators and flags in every packet header. After two hosts establish a Telnet session, there is no need to resend this information. For example, sending just three characters in a TCP/IP session requires 43 bytes to be transferred (each IP and TCP header uses 20 bytes). By using CSLIP, the 40 bytes of header overhead can be reduced to between 3 and 5 bytes. Information such as IP addresses, TCP fragmentation indicators, and type of service (ToS) flags need not be sent, as illustrated on the Compressed SLIP Diagram.

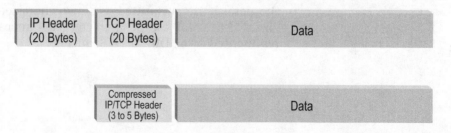

Compressed SLIP

PPP

When a more robust and flexible serial protocol is required, the TCP/IP community relies on PPP. This Internet standard protocol offers multiprotocol support, data compression, host configuration, and link setup.

PPP is based on the HDLC standard, which deals with LAN and WAN links and operates at the Data Link Layer of the OSI model. PPP starts with the HDLC frame format, but adds a protocol field to identify the Network Layer protocol of each frame. This is what allows a PPP link to carry data for multiple network protocols.

PPP is used by higher-layer protocols, such as TCP/IP, to provide simple WAN connectivity between users. It replaces SLIP and solves some of the inefficiencies found in SLIP. PPP supports either asynchronous (character-oriented) or synchronous (bit-oriented) transmission links.

There are three PPP frame formats, depending on what it is carrying:

- Information frames
- Link control frames
- Network control frames

Information Frames

The PPP Information Frame Diagram presents the fields in a PPP information frame header.

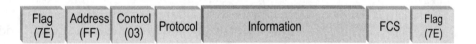

PPP Information Frame

- **Flag**—1 byte—Used for synchronizing the bit stream—"7E."
- **Address**—1 byte— The address field is always "FF."
- **Control**—1 byte—The control field is set to "03."
- **Protocol**—2 bytes—The protocol field contains addressing for the higher layers. This field is similar (but not identical) to the Ethernet type field (Ethertype). Some common addresses are:
  - 0021H—TCP/IP
  - 0023H—OSI

331

    –   0027H—Digital Equipment Corporation (DEC)

    –   002BH—Novell

- **Information**—variable—The information field contains data that may be preceded by Network Layer headers such as IP.

- **Frame Check Sequence (FCS)**—2 bytes—The FCS field is used to ensure data integrity.

- **Flag**—1 byte—The flag field signals the end of a frame, and possibly the start of the next frame. It is set to "7E."

Link Control Frames

A link control protocol (LCP) can be used to specify certain data link options, such as which characters are going to be released on an asynchronous link. It can also be used to negotiate not having to send a flag or address byte, and reduce the size of the protocol field from 2 bytes to 1 byte for more efficient line use. The PPP Link Control Frame Diagram presents an LCP frame header.

PPP Link Control Frame

Network Control Frames

The third format of a PPP frame is used to negotiate variables such as using header compression. The PPP Network Control Frame Diagram presents a frame used for this purpose. This protocol can also be used to dynamically negotiate the IP addresses for each end of the link, such as when you connect to your ISP over a dial-up connection.

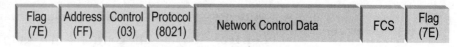

PPP Network Control Frame

Zero Bit Insertion   As in HDLC frames, PPP frames require a special method for sending a flag byte, or control character, as part of a frame's data. On a synchronous link, this is taken care of by the hardware using a technique called zero bit insertion. The American Standard Code for Information Interchange (ASCII) escape character ("7D") is sent first, followed by the data character with its sixth bit complemented (reversed). In other words, the sixth bit of a data character is reversed, so it does not appear to be a control character. The "7D" code just before the altered byte alerts the receiver that the character has been changed.

For example, if the flag character "7E" appears as data on an asynchronous link, it is sent as a 2-byte sequence of "7D" "5E," because "7E" (01111110), with its sixth bit complemented, is equal to "5E" (01011110). If the escape character itself must appear in the data, it would be sent as "7D" "5D." In addition, ASCII control characters (any value lower than a "20") would be sent the same way. For example, a Bell (BEL) character, which causes a personal computer (PC) speaker to beep, is a hexadecimal "07" (00000111). It would be sent as a 2-byte sequence "7D" "27" (00100111).

Activities

1. List the properties of PPP that provide a more robust service than SLIP.

   _____

   _____

   _____

   _____

   _____

   _____

2. Match the feature with the supporting protocol.

   PPP

   CSLIP

   SLIP

   a. Reduces the TCP header size _____

   b. Dynamically negotiates IP addresses _____

   c. Can only carry one type of higher layer protocol at a time _____

   d. Frame similar to HDLC frame _____

   e. Provides no checksum _____

   f. Supports either asynchronous or synchronous links _____

Extended Activity

Download and review RFC 1144 from the Internet.

Lesson 3—Integrated Services Digital Network

Having discussed WAN protocols at the Physical and Data Link Layers, we turn our attention to protocols that move information across switched networks. We will discuss ISDN first, because it is widely implemented and provides a framework for understanding other WAN products that provide integration of multiple services.

Objectives

At the end of this lesson you will be able to:

- Describe ISDN protocols and services
- Describe the differences between Basic and Primary Rate ISDN services

Key Point

ISDN provides transport of multiple user services.

Integrated Digital Networks

As analog transmission and switching components were rendered obsolete by superior digital technologies, a new set of protocols was needed to benefit from the full potential of the digital systems. ISDN provides a framework for the development of these components and protocols.

We have already seen how telephone companies have largely completed conversion of their voice networks from analog to digital. We have also seen how they have made the resulting integrated digital network (IDN) directly accessible for data communications by first offering Digital Data Service (DDS) and, later, T-carrier service in North America (and equivalent services elsewhere). We have also seen the limitations of these services.

ISDN represents a logical migration of voice-oriented IDN toward a network that serves multiple purposes (voice, data, video, fax, and all other forms of electronic communication), regardless of the source. The ISDN Network Diagram illustrates the services ISDN can transport.

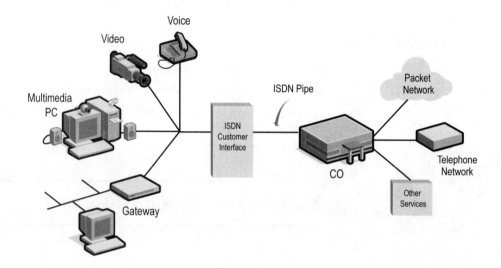

ISDN Network

ISDN can be characterized in two ways:

- As a bundle of services offered for transmission of voice, data, and other forms of communication by means of the switched telephone networks of the world

- As a set of protocols that defines a standard interface to the network, making it possible for many vendors to supply both hardware and software to take advantage of the services offered

This lesson considers ISDN a bundle of services.

ISDN Services

ISDN service is provided by telephone companies, thus we subscribe to ISDN in the same way we subscribe to voice services. Of course, to get ISDN service, the nearest telephone company central office (CO) must offer it. Only some COs currently offer this technology, because the process of making ISDN available is a gradual one.

Public-switched telephone networks (PSTNs) have largely adopted common-channel signaling (CCS) using the digital Signaling System 7 (SS7) signaling protocol. ISDN takes advantage of this capability. SS7 is an international high-speed, packet-switching protocol used on telecommunications backbones that provides redundant data and signaling paths between telephone switching offices. All CCS for SS7 and ISDN is performed out-of-band, that is, on separate channels than those used for voice conversations.

ISDN provides three types of channels for subscriber use, as listed in the ISDN Access Table:

*   **D channels**—These channels are provided for CCS control signaling; however, they can also be used for data. D channels operate at 16 kilobits per second (Kbps). Each D channel is associated with one or more channels of another type. The channel can be used, for example, to tell a telephone company to which ISDN subscriber the other channels are to be connected. CCS eliminates the problem of distinguishing signals from data. By using a single channel for signaling for several data channels, bandwidth is saved. A D channel can also be used for transmitting certain types of data that require low bit rates.

*   **B channels**—These "bearer" channels operate at 64 Kbps, and are used for data, voice, fax, slow-scan video, and so on. Slow-scan video refers to video applications that do not require smooth motion of pictures, such as transmitting the slides of a presentation.

*   **H channels**—These channels operate at 384 (H0), 1,536 (H1), or 1,920 (H2) Kbps, and are used for applications requiring high bandwidth, such as backbone networks and full-motion video. They can also be multiplexed by a subscriber in the same manner as T-carrier channels.

ISDN Access

Channel	Rate (Kbps)	Applications
D	16	Control signaling
B	64	Data Voice Facsimile Slow-scan video

ISDN Access

Channel	Rate (Kbps)	Applications
H0	384	Backbone networks
H1	1,536	Full-motion video
H2	1,920	Multiplexing

These access speeds are often referred to as Narrowband ISDN. Broadband ISDN (B-ISDN) is ISDN at speeds of 154 Mbps and higher.

Note that the basic building block for ISDN channels is Digital Signal Level 0 (DS0):

- B channel = 1 x DS0

- H0 channel = 6 x DS0

- H1channel = 24 x DS0

- H2 channel = 30 x DS0

The D channel is used for control, and typically does not carry data.

We may choose between two ISDN services, depending on our needs. Of course, we might require more than one kind of service, just as we might require more than one telephone line. The basic services include:

- **Basic rate service**—Also known as Basic Rate Interface (BRI), this service provides one D channel and two B channels and is sometimes referred to as "two B plus D" or "2B+D." This service provides 144 Kbps of usable capacity; however, framing, synchronization, and other overhead reduces the total bit rate of BRI service to 128 Kbps.

- **Primary rate service**—Also known as Primary Rate Interface (PRI), this service is structured around the bandwidths of T1 for North America (1.544 Mbps) and E1 elsewhere (2.048 Mbps), including Japan. It includes an optional D channel and a number of B and H channels, in combinations that do not exceed the allowable bandwidth when necessary overhead is included. In North America, PRI service provides 23 B channels and 1 D channel, or "23B+1D."

ISDN-PRI service can be an economical alternative to a group of single analog trunks. Depending on local pricing and implementation, ISDN-PRI usually becomes cost-effective when a business requires 11 to 25 trunks.

After selecting the required services, connections for ISDN channels are set up in several ways:

- **Semipermanent**—Set up by prior arrangement, this is the ISDN equivalent of a leased line.

- **Circuit-switched**—This is similar to using a modem on today's public-switched network to establish a connection to another user. An important difference is that a D channel is used to transmit the control information necessary to establish and terminate a call.

- **Packet-switched**—This is X.25 packet switching. Software is available that allows ISDN networks to appear as if they were X.25 networks, performing packet switching at the Network Layer.

ISDN also provides a number of services not previously available to data communications users. Many of these are similar to services provided to users of the voice network, because they are both made possible by the same SS7 signaling protocol. For example, ISDN automatically provides a caller's telephone number as part of incoming calls. This feature, called Automatic Number Identification (ANI), is similar to Caller ID, but cannot be blocked by the caller.

With ISDN, a system can also block incoming calls, transfer a call to another ISDN subscriber, and connect to multiple ISDN subscribers (like a conference call).

Call-by-Call Service

In non-ISDN telephone systems, each trunk must be permanently assigned to a particular ToS. For example, certain trunks are designated for foreign exchange connections, tie trunks, incoming or outgoing calls, or data connections. Because the communication patterns of most businesses change frequently through the day, this "hard-wired" approach can be very inefficient.

In contrast, with ISDN-PRI call-by-call service, any B channel can be configured to support any communications activity on an as-needed basis. Thus, if a business suddenly receives a flood of incoming calls, B channels are automatically assigned to handle the load. Or, if the company transfers large volumes of data after hours, the B channels that handle voice during the day can carry data at night.

Channel Bonding and NFAS

Like T1 channels, multiple 64-Kbps ISDN channels can be bonded, or combined to form a single large channel. ISDN channel bonding is performed by the D channel as part of its control signaling function, or through the use of a higher-layer protocol, such as Point-to-Point Multilink Protocol (MP).

Non-facility associated signaling (NFAS) allows customers to convert surplus ISDN-PRI D channels to B channels. Each D channel can control the B channels of several PRI lines. For example, a customer with two ISDN-PRI lines can designate one D channel to control the B channels of both circuits. The unused D channel can then be converted to an additional B channel. This approach provides a total of 47 channels for voice or data, rather than the 46 channels possible without using NFAS.

ISDN Signaling

As we discussed earlier, ISDN makes the rich features of SS7 available to corporate and individual subscribers. To deliver that service to the desktop, two additional families of protocols are necessary, as illustrated on the ISDN Signaling Protocols Diagram:

- Digital Subscriber Signaling System No. 1 (DSS1) controls ISDN signaling between the CO switch and subscriber's premises.

- Private Signaling System No. 1 (PSS1, also called Q.SIG) controls ISDN signaling within a network of private branch exchange (PBX) switches.

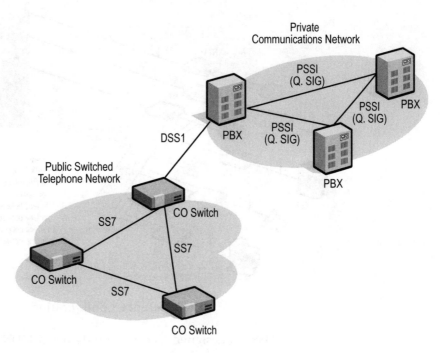

ISDN Signaling Protocols

DSS1

Digital Signaling System No. 1 (DSS1) allows an intelligent user terminal to access ISDN network services. The user terminal may be an individual ISDN telephone, single computer, or PBX.

DSS1 operates at the S and T interfaces of the ISDN reference configuration, as illustrated on the ISDN Reference Configuration Diagram. It carries ISDN service to a terminal adapter that converts ISDN to analog, or ISDN-capable, terminal equipment that uses the ISDN signal directly.

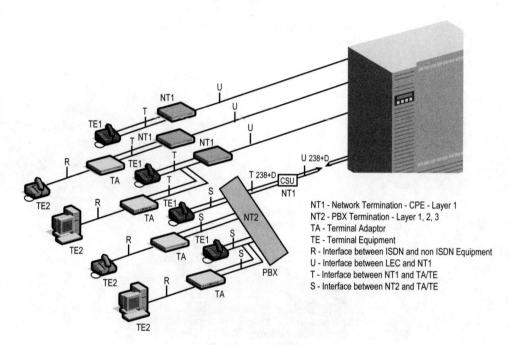

ISDN Reference Configuration

DSS1 is designed to carry standardized ISDN transmission to the customer's premises, but not within them. If the ISDN service is directly connected to a few receiving devices, DSS1 is all that is necessary to deliver ISDN to the customer. The end device either uses the signal directly, or an ISDN-capable PBX completes the service to a directly connected user.

However, receiving ISDN service is more complex for customers with private communications networks, composed of multiple PBX systems. To deliver ISDN features, such as calling party identification, across a private network, the PBXs must support the transmission of ISDN signaling.

This is not necessarily a problem, because most major switch models support ISDN. However, each PBX manufacturer has traditionally used proprietary standards for signaling between its switches. As customer companies and PBX vendors have merged and reorganized, many private networks now include incompatible PBXs from several vendors. To enable seamless ISDN across these mixed networks, a new ISDN protocol was required: PSS1.

PSS1

Private Signaling System No. 1 (PSS1), or Q.SIG, is an international standard created and maintained by several groups, including the European Telecommunications Standards Institute (ETSI), European Computer Manufacturers Association (ECMA), and International Organization for Standardization (ISO).

By using Q.SIG signaling between its switches, a company can provide ISDN supplemental services across a private network of PBX systems from multiple vendors. For example, Q.SIG makes it possible to provide services such as call forwarding, call waiting, or calling party identification across a tandem network of mixed switching systems.

PSS1 is commonly known as Q.SIG because it handles signaling at the Q reference point in the ISDN reference configuration. This reference point was created, along with the protocol, to describe the logical communication between switches in a private network.

Unlike other reference points in the ISDN reference configuration, the Q point describes the logical communication function, not the physical interface to the network. Physical connections between switches are described by the C reference point, as illustrated on the Q and C Reference Points Diagram. The Q.SIG specification makes this distinction because a single logical connection may operate over one or more physical connections.

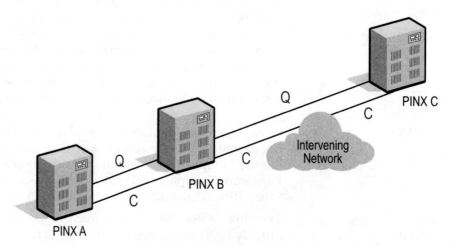

Q and C Reference Points

The Q.SIG specification defines each node in a private ISDN network (PISN) as a private ISDN network exchange (PINX). All PINXs in a PISN operate as peers, using symmetrical communication; in other words, there is no "user side" or "network side" to a Q.SIG communication.

Q.SIG and ISDN Supplementary Services

Because Q.SIG is based on the underlying standards of ISDN, it supports all ISDN supplemental services: call transfer, call forwarding, calling party identification, and so on. This compatibility allows Q.SIG to enable ISDN features across a private network. Q.SIG also specifies many supplementary services in addition to those offered by ISDN.

The most important Q.SIG supplementary service, Generic Functions, allows each switch vendor to offer additional nonstandard services and features. To enable a nonstandard service, only the sending and receiving switches must support it; any intermediate switches simply transmit the signaling information transparently.

For example, consider a private communications network composed of PBX systems from two vendors. Vendor A has implemented value-added services not supported by Vendor B. However, both vendors' products support Q.SIG. Therefore, a user attached to a Vendor A switch can invoke a value-added service of Vendor A when calling a user connected to another Vendor A switch. The switches from Vendor B do not need to support the service; they merely forward the signaling data without trying to interpret it.

This compromise gives both vendors and customers what they need and want. Vendors can comply with the standards customers demand, without sacrificing the opportunity to offer competitive features. Customers can select preferred features from several vendors, with the assurance that any PBX will support the common ISDN and Q.SIG feature set.

Activities

1. Research whether ISDN is offered in your area, and the rate charged for the service.

   _____

   _____

2. Discuss the capabilities of ISDN, and whether you think ISDN will be available for a long time as a service.

   _____

   _____

   _____

   _____

3. Discuss the concept of "integrated services," and what this means today vs. when ISDN was originally developed.

   _____

   _____

   _____

   _____

Extended Activity

1. Go to the following Web sites and research ISDN information.

   a. ISDN Tutorial:
      http://public.pacbell.net/ISDN/connect.html

   b. ISDN Information and Web Links:
      http://webopedia.internet.com/TERM/I/ISDN.html

Lesson 4—Asynchronous Transfer Mode

ATM is a mature technology developed to transmit voice, video, and data across LANs, MANs, and WANs ATM is an international standard defined by ANSI and International Telecommunication Union-Telecommunications Standardization Sector (ITU-T). ATM implements a high-speed, connection-oriented, cell-switching, and multiplexing technology designed to provide users with virtually unlimited bandwidth.

ATM combines the best features of two transmission methods. Its connection-oriented nature makes ATM a reliable service for delay-sensitive applications such as voice, video, or multimedia. Its flexible and efficient cell switching provides quick transfer of other forms of data.

Many telecommunications carriers have built their nationwide data networks upon ATM technology. Every major data networking equipment vendor offers a line of ATM products.

Objectives

At the end of this lesson you will be able to:

- Detail how ATM networks are constructed
- Describe how an ATM network transfers information from source to destination
- Identify the basic characteristics of the ATM protocol
- Describe the functions of the fields of an ATM protocol header

 **Key Point**

ATM provides high-speed, connection-oriented, cell-switching services.

The Need for ATM

In the mid-1980s, telecommunications researchers began to investigate technologies that would serve as the basis for the next generation of high-speed voice, video, and data networks. The result of this research was the development of B-ISDN standards.

B-ISDN was designed to support subscriber services that require both constant and variable bit rates. These services include data, voice, video, imaging, and multimedia applications. The ultimate goal of B-ISDN is to replace the current public network infrastructure and become the universal network of the future. ATM is the foundation on which B-ISDN is to be built.

Key Point

ATM combines the strengths of STM and PTM.

Transfer Modes

A transfer mode specifies a method of transmitting, multiplexing, and switching data in a network. Three transfer modes were considered as possible candidates for B-ISDN, including:

- Synchronous Transfer Mode (STM)
- Packet Transfer Mode (PTM)
- ATM

STM

"Synchronous" means that data communication is organized by a microprocessor clock; a receiving node can detect the beginning and end of a signal because signals start and stop at particular times. Networks that use STM divide each transmission frame into a series of time slots, and then allocate particular time slots to each user. For example, on the Synchronous Transfer Mode Diagram, Time Slot 2 is dedicated to the same user in each and every frame.

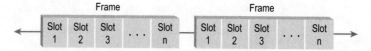

Synchronous Transfer Mode

STM is ideal for transmission of voice and video, because it provides a constant-bit rate service. Voice and video require predictable and guaranteed network access, or the quality of the transmission degrades rapidly.

In contrast, data transmissions are typically bursty; a user is idle for relatively long periods of time between short periods of intense data-transfer activity. For example, in a typical client-server transaction, the client request consumes very little bandwidth. However, when the server responds, a large amount of data is typically transmitted from the server back to the client. The server's response may actually consume the entire bandwidth of the network.

STM is inefficient for data communications because the same time slot in each frame is reserved for a particular user, regardless of whether the user has data to transmit. When a user is idle, the time slot is wasted, because STM does not reassign unused time slots to other users.

Examples of STM technologies include standard T1 circuits, E1 circuits, and Synchronous Digital Hierarchy (SDH) circuits provided by telecommunications carriers.

PTM

In a network technology based on PTM, data is broken into variable-size units of data (packets, datagrams, or frames). Each unit contains both user data and a header that provides information for routing, flow control, and error correction. Instead of establishing a dedicated physical connection between the source and destination station, the network relays packets from one node to another, often in multiple parallel paths, until they reach their final destination. PTM is implemented through technologies such as Ethernet, Token Ring, Fiber Distributed Data Interface (FDDI), X.25, and frame relay. The Packet Transfer Mode Diagram illustrates the concept of PTM.

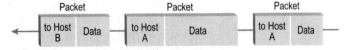

Packet Transfer Mode

PTM is excellent for bursty data applications because a station only consumes bandwidth when it needs to transmit data. When a station is idle, its share of network bandwidth can be used by other stations. The resulting variable transmission rates and delay, within reasonable limits, are not critical issues for data communications.

However, PTM does not provide the guaranteed network access required by constant-bit rate applications such as voice or video. Voice and video tolerate very little delay in transmission; however, they can handle some loss or inaccurate information.

ATM

We have seen that STM is excellent for voice and video applications, but it is inefficient for data applications. On the other hand, PTM is excellent for data applications, but cannot provide the guaranteed bandwidth and low delay required for voice and video.

ATM offers the best of both worlds. It combines the strengths of STM (constant transmission delay and guaranteed capacity) and PTM (flexibility and ability to handle intermittent traffic) in a single transfer mode that meets the needs of voice, video, and data applications.

In computing, the term "asynchronous" usually means that data transmission is coordinated through start and stop signals, without the use of a common clock. However, ATM networks use "asynchronous" to describe how network bandwidth is assigned to user applications. ATM assigns network access to users based on demand, which means that locations in the synchronous data stream are assigned to users in a random, or asynchronous, pattern. The ATM Diagram illustrates the concept of ATM.

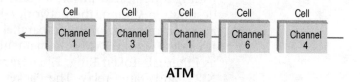

ATM

The ultimate goal of ATM is to provide extremely high-speed communications that allow voice, video, and data applications to run across a single integrated network. By combining connection-oriented transmission with the use of small, fixed-length cells, ATM solves many of the problems encountered when these applications share the same network.

- The connection-oriented nature of ATM provides minimal delays for voice and video applications.

- The use of fixed-length cells simplifies switch design by allowing the switch logic to be implemented in silicon, in other words, in the switch firmware instead of in software. This greatly reduces the processing time required for each cell, increases switch throughput, and reduces the cost of the switching technology.

- Small video cells are not delayed by large data cells because all cells are the same size. This means it is relatively easy to predict the amount of network delay between any two points. In addition, the variation in delay is significantly decreased, because time-sensitive applications such as voice and video can share the same transmission facilities with data applications.

- The use of fixed-length cells rather than time slots overcomes the major weakness of STM: wasted bandwidth. An ATM station only consumes bandwidth when it has data to transmit.

Connection-Oriented Mode

In a connectionless network, a predefined end-to-end connection between source and destination nodes is not required for data transmission. As a result, data flow across the network occurs along the best available path, rather than over a predefined path. A connectionless service is sometimes referred to as a datagram service. IP is an example of a connectionless service.

ATM operates in a connection-oriented mode. In an ATM network, a pair of source and destination nodes establishes a virtual connection before the source begins transmitting data. All cells transmitted between a pair of source and destination nodes follow the same virtual connection, or virtual path (VP), through a network of ATM switches during the transmission. A later transmission between the same source and destination may follow a different VP, but the path will not change for the duration of the transmission. This approach improves overall transfer speed by making it simpler and faster to switch cells through intermediate nodes.

A connection-oriented service requires that a virtual connection be established between source and destination nodes before data can be transmitted. The ATM and Virtual Circuits Diagram illustrates the relationship between these two concepts. As previously mentioned, all switched virtual circuit (SVC) connections involve three phases:

- Connection establishment
- Data transfer
- Connection termination

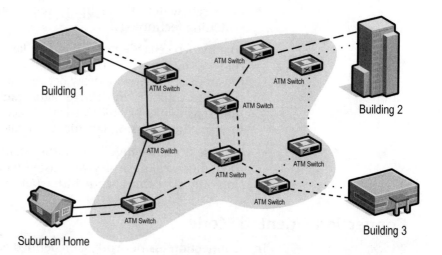

ATM and Virtual Circuits

Connection-oriented networks trade greater complexity required in end nodes to support signaling and connection setup, in return for much greater simplicity in intermediate (switching) nodes. A connection-oriented network has several advantages over a connectionless network when attempting to support real-time, high-speed applications.

A connection-oriented network enables a network to guarantee a minimum level of service. If a network does not have sufficient resources to accept a connection request, the network simply refuses to establish the connection. This guarantees the network will have sufficient resources to support all active connections, and that queue overflows will not occur.

A logical connection between users means the signals travel over the same logical path for the duration of the connection, and switching delay is virtually eliminated. This is important because both voice and video applications are extremely sensitive to variations in transmission delay.

The devices at each end of a virtual circuit may operate at different speeds, because an end-to-end physical connection is not established. This allows data to be transmitted at one speed by the source node, while it is received at a different speed by the destination node.

Use of the connection is relatively high while the connection is established. When there is no more data to be transmitted, the connection is terminated, and the previously allocated network resources may now be used by another connection.

Fixed-Length Data Cells

ATM organizes transmission by formatting data into fixed-size units called cells. Each cell contains 53 bytes that are divided into a 48-byte payload (data) field and a 5-byte header. The ATM Cell Diagram illustrates the basic format of an ATM cell.

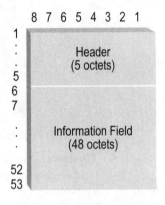

ATM Cell

The use of fixed-length cells simplifies switch design by enabling a switch to be implemented in silicon. In other words, the switching logic is implemented in hardware devices without the need for an operating system, as in routers. This greatly reduces the processing time required for each cell, increases switch throughput, and reduces the cost of the switching technology.

ATM's cell-switching approach also makes efficient use of network bandwidth for bursty data transfers, by allocating cells to applications only as needed.

Header Details

Each ATM cell's 5-byte header contains six fields, as illustrated on the ATM Header Diagram.

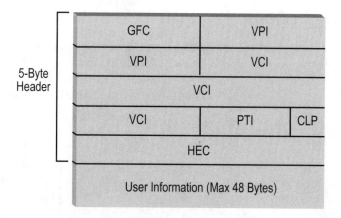

ATM Header

The six fields of the ATM header include:

- **General flow control (GFC)**—GFC is 4 bits long and is reserved for future flow control use between a user and network.

- **Virtual path identifier (VPI)**—VPI is 1 byte long and identifies the VP to which a cell belongs. The VPI has local significance, and each ATM device may change the VPI as the cell traverses the network to its destination.

- **Virtual channel identifier (VCI)**—VCI identifies the VC to which a cell belongs. The VCI is a 2-byte field. As with the VPI, the VCI also has local significance, and may change as the cell traverses the network. The VPI/VCI pair tell the ATM switch how to switch a cell.

- **Payload type indicator (PTI)**—PTI is 3 bits long. The PTI indicates whether the payload contains user, control, or management data. The PTI field also indicates whether the cell has encountered congestion as it traverses the ATM network.

- **Cell loss priority (CLP)**—This single bit is set to 1 to indicate a cell with a low priority; the default is 0, which indicates a regular priority cell. In the event of network congestion, an ATM switch discards low priority cells (CLP = 1) before discarding cells with a regular priority (CLP = 0). The CLP function is important because it allows certain types of traffic to take priority in a congested network.

- **Header error control (HEC)**—These 8 bits provide a cyclic redundancy check (CRC) to detect errors in the cell header. Its main function is to validate the VPI and VCI fields to protect against the delivery of cells to the wrong User-Network Interface (UNI). Although this field is transmitted as part of the cell header, it is computed and used by the Physical Layer, not the ATM layer.

B-ISDN and ATM

ATM is a subset of the overall B-ISDN standard. The B-ISDN protocol stack is presented on the B-ISDN Protocol Stack Diagram. The three primary layers that we will discuss are the bottom three layers. In general, these layers correspond to the Physical and Data Link Layers of the OSI model, and include:

- ATM adaptation layer (AAL)

- ATM layer

- Physical Layer

These three layers are all contained in one protocol header.

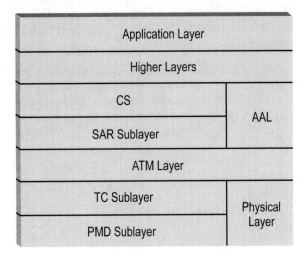

B-ISDN Protocol Stack

AAL

AAL is an end-to-end process used only by the two communicating end nodes to insert and remove data from the ATM layer. AAL maps higher layer services, such as TCP/IP or video services, to the ATM layer for transport across the network. AAL provides five different protocols, AAL 1 through AAL 5. Each AAL protocol consists of a specific convergence sublayer (CS) and specific SAR sublayer.

The most popular data communications protocols, such as IP, NetWare, and AppleTalk, make use of variable-length packets. These packets are almost always larger than can be carried in the payload field of a single ATM cell. For most ATM devices to exchange data, they must make use of an adaptation layer protocol that segments higher-layer protocol packets into cells for transmission across the network, and reassembles cells received from the network into the original data packet.

AAL 1, constant bit rate (CBR), is designed for CBR data that requires synchronization between sender and receiver. This service is used by voice, video, and similar traffic. AAL 1 adds 4 bytes of overhead for encoded timing information to address the close coordination needs of CBR traffic. Typical protocols used by AAL 1 are DS0s, DS1s, and DS3s, allowing an ATM network to emulate voice or DS-type services.

AAL 2 addresses the needs of variable bit rate (VBR) services. The additional overhead needs for AAL 2 are not yet defined.

AALs 3 and 4 are designed for connection-oriented and connectionless data services, respectively. AALs 3 and 4 traffic includes X.25, frame relay, and ISDN D channel signaling.

AAL 5, simple and efficient adaptation layer (SEAL), is used for VBR data transmitted between two users over a preestablished ATM connection. SEAL assumes that higher-layer processes will handle error recovery. It is the only AAL process that does not add any overhead to the user information sent to the lower levels. Each AAL process is performed at the CS.

The purpose of the two AAL sublayers is to convert the user data into 48-byte data fields for transmission in an ATM cell. The unit of data produced at this layer is a protocol data unit (PDU).

Convergence Sublayer (CS)

The CS is service-dependent. Services provided by the CS include: handling cell delay variation, source clock recovery, monitoring lost and misinserted cells and possible corrective action, monitoring user information for bit errors, and reporting on the status of end-to-end performance. In addition to these services, CS is also responsible for dividing data from the next higher layer into logical packet data units (CS-PDU) that are usable by the segmentation and reassembly (SAR) sublayer. The processes associated with AAL, and its sublayers CS and SAR, are illustrated on the AAL Process Diagram.

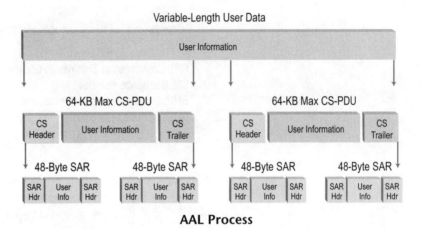

AAL Process

Segmentation and Reassembly (SAR) Sublayer

The SAR sublayer is responsible for segmentation of higher-layer information into 48-byte fields suitable for the payload of ATM cells, and the inverse operation; reassembly of cell payload information for the next higher layer. In AAL processes AAL 1 through AAL 4, the SAR sublayer adds data to the PDU sent to the ATM layer. No SAR structure is shown for AAL 5, because that protocol does not add overhead at the SAR sublayer. The SAR-PDUs are presented on the AAL Structure Diagram.

SAR Structure for AAL 1

CSI: Convergence Sublayer Indication
SN: Sequence Number
SNP: Sequence Number Protection
SDU: Service Data Unit

SAR Structure for AAL 2

SN: Sequence Number
IT: Information Type
LI: Length Indicator
CRC: Cyclic Redundancy Check

SAR Structure for AALs 3 and 4

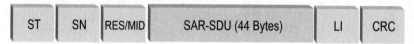

ST: Segment Type
RES: Reserved
MID: Multiplexing Identifier

AAL Structure

ATM Layer

The ATM layer is responsible for data transmission between adjacent nodes, and transporting data between AAL and the Physical Layer. During transport of data between AAL and Physical Layers, the ATM layer adds (or removes) 5 bytes of header information to the SAR-PDU, making a complete ATM cell. In addition to transporting data, the ATM layer is responsible for multiplexing and demultiplexing cells into a single cell stream on the Physical Layer. The ATM Layers and Routing Diagram illustrates the relationship between these two concepts.

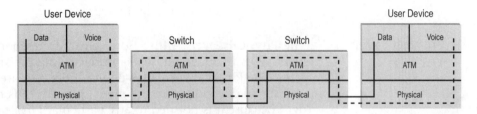

ATM Layers and Routing

The major function of the 5-byte ATM header is to identify the virtual connection to which the cell belongs. The traditional functions of packet headers, such as sequence numbers for error correction and flow control and checksums, are eliminated. This implies that an ATM switch can process a cell very quickly, which eliminates queuing delays while increasing switch throughput.

Each ATM cell contains a small 48-byte information field. The use of this small field reduces the number of internal buffers a switch must support, while reducing the queuing delay for the switch's buffers. This enables an ATM switch to process cells very quickly, which reduces latency and increases throughput. The switch does not have to allow for varying frame sizes.

Unlike an X.25 network, ATM does not support error correction or flow control on a link-by-link basis. This means that if a physical link introduces a bit error, or is temporarily overloaded resulting in the loss of a cell, no corrective action is taken. An ATM network does not support a facility that allows the node at one end of a point-to-point physical link to request the retransmission of lost or corrupted cells from the node at the other end of the physical link.

There are two reasons it is unnecessary for an ATM network to support these functions:

- The introduction of fiber-based digital transmission facilities has created a relatively error-free transmission environment. Fewer transmission errors mean there is a reduced need for a network to perform error correction.

- Workstations run higher-level protocols, such as TCP, that perform error correction and retransmission. Because the ATM network can concentrate on just switching cells and not worry about error correction, cell throughput at each switching node is substantially increased.

Physical Layer

The Physical Layer defines how cells are transported over a network. This includes physical interfaces, media, and information rates. The Physical Layer also defines how cells are converted to a line signal depending on the media type. ATM is media-independent because it is not tied to any particular Physical Layer.

The Physical Layer consists of two sublayers:

- Transmission convergence (TC) sublayer
- Physical medium-dependent sublayer

TC Sublayer

The TC sublayer converts between ATM cells and the bit stream clocked to the physical medium. On transmission, TC maps the cells into the time-division multiplexing (TDM) frame format (or the appropriate underlying physical transport protocol). On reception, TC must delineate the individual cells in the received bit stream, either from the TDM frame directly, or by means of the HEC in the ATM cell header.

The HEC code is capable of correcting any single-bit error in the header. It is also capable of detecting many patterns of multiple-bit errors. The TC sublayer generates HEC on transmission, and uses it to determine whether the received header has any errors. If errors are detected in the header, the received cell is discarded. Because the header tells the ATM layer what to do with the cell, it is very important that the header be free of any errors. Otherwise, the cell might be delivered to the wrong user, or an undesired function in the ATM layer may inadvertently be invoked.

The TC sublayer also uses HEC to locate cells mapped into a TDM payload. The HEC will only match data in the 5-byte header, not in the payload. Thus, the HEC can be used to find cells in a received bit stream. After several cell headers have been located through the use of HEC, TC knows to expect the next cell 53 bytes later. This process is referred to as HEC-based cell delineation.

TC Cell Rate Decoupling

The TC sublayer also performs cell rate decoupling (or speed matching function). Physical media that have synchronous cell time slots such as T-carriers, Synchronous Optical Network (SONET), and SDH require this function; asynchronous media such as FDDI do not.

Special codings in the ATM cell header indicate whether a cell is either unassigned or idle. The transmitting switch multiplexes multiple VPI/VCI cell streams, queuing them if an ATM slot is not immediately available. If the queue is empty when the time comes to fill the next synchronous cell time slot, then the TC sublayer inserts an unassigned or idle cell to maintain a constant bit stream. The receiver discards unassigned or idle cells and distributes the other, assigned cells to their destinations.

Physical Medium-Dependent Sublayer

This lowest layer on the B-ISDN protocol stack is the only fully medium-dependent layer. This sublayer must guarantee proper bit timing reconstruction at the receiver. Therefore, it is the responsibility of the transmitting peer entity to provide for insertion of the required bit timing information and line coding appropriate to each type of physical medium.

Activities

1. Describe what happens when an ATM network transfers information from source to destination.

   _____

   _____

   _____

   _____

   _____

   _____

2. What are the functions of the fields of an ATM protocol header?

   _____

   _____

   _____

   _____

   _____

   _____

Extended Activity

Visit the ATM Web site **http://www.atmforum.com** and review the white papers provided regarding ATM technology. Review the products mentioned in the text.

Lesson 5—Frame Relay

Frame relay is an adaptation of the ISDN interface for the purpose of moving multiplexed data across a WAN. This standard defines an interface between an enterprise network and packet-switching network. The term frame is used because frame relay builds data frames and can asynchronously multiplex these frames from multiple virtual circuits (endpoints) into a single high-speed data stream. The term relay is used because each frame relay device forwards frames as they move through the network without examining the frame's payload or demultiplexing the data stream. Most local and long distance carriers offer frame relay.

This lesson evaluates frame relay from a technical perspective, focusing on how the available bandwidth is allocated and how virtual circuits are created in a frame relay network. This lesson also reviews the protocol details of frame relay.

Objectives

At the end of this lesson you will be able to:

- Describe the basic operation of frame relay
- Describe how frame relay uses virtual circuits to move data across a network
- Draw a frame relay frame encapsulating higher layers
- Understand the protocol header used by frame relay
- Describe how frame relay is typically implemented in an enterprise network
- Understand the difference between public, private, and hybrid frame relay networks
- Evaluate the advantages and disadvantages of frame relay vs. private line networking

 **Key Point**

Frame relay uses virtual circuits to move data across a WAN.

What is Frame Relay?

A frame relay network is a switched network that moves data frames from one network to another network. Logically, a frame relay is an electronic switch or switches running frame relay software. Physically, it is a box that connects to three or more high-speed links, and routes data traffic between them. The Frame Relay Diagram illustrates this operation. Suppose virtual circuits have been established as shown on the diagram: Virtual Circuit 1 (VC1) from Multiplexer (MUX) A to MUX C, Virtual Circuit 2 (VC2) from MUX A to MUX D, and Virtual Circuit 3 (VC3) from MUX A to MUX E. All three circuits flow through Frame Relay B.

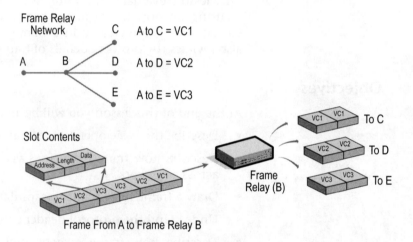

Frame Relay

Next, suppose data for all three circuits flows into MUX A. The MUX places each circuit's data into frames, storing an address and length with each frame (this diagram was simplified by showing all data the same length).

The frames from all three circuits are then transmitted from MUX A to MUX B. MUX B then demultiplexes the frames, and creates new frames to transport the data to MUXs C, D, and E.

Another view of a similar network is shown on the Frame Relay Network Diagram. Here, routers feed a frame relay network. VC1 could be a path from Router 1 to Router C. VC2 could be a path from Router 1 to Router D, and VC3 could be a path from Router 2 to Router E.

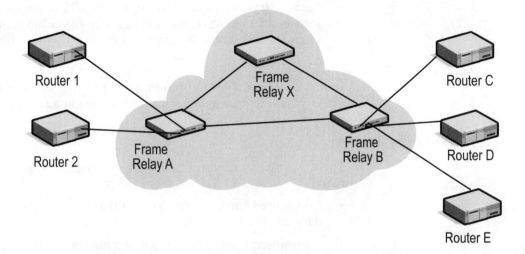

Frame Relay Network

Frame relay has several characteristics including:

- As defined by ITU-T, frame relay can be T1 or E1 bandwidth. It is ITU-T's intention that frame relay fall under cell relay (ATM), with respect to satisfying user bandwidth requirements. However, current frame relay service providers are promising T3 bandwidth; thus, frame relay will overlap with cell relay at its high end.

- Frame relay was originally intended for data communications, rather than voice, video, or other time-sensitive information.

- Only connection-oriented service is provided. Unlike other ISDN packet services, the network does not offer a complete connection mode data link service. Although frame relay is connection oriented, it does not provide for end-to-end error detection and correction; a frame either makes it across the network or it does not. The frame's address field and CRC are local to each interface and are changed by the network as each frame moves through the network devices.

- Frame relay discards frames with errors without providing notification to the sender or receiver; it assumes the physical links are reliable. Error recovery is left up to higher-layer protocols, such as TCP/IP.

- Frame relay, a Data Link Layer protocol suite, is faster than X.25, a Network Layer suite. Whereas X.25 provides error correction and recovery, windowing, and other higher-layer services, frame relay only delivers frames to the next frame relay device.

Frame relay's variable-length frames are not well suited for voice or video, which require a data stream that flows at a predictable rate.

Frame Relay Terms

There are two important terms to understand regarding frame relay:

- Committed information rate (CIR) is the guaranteed average data rate for a particular service.

- Committed burst size (CBS) is the number of bits that can be transferred during some time interval.

The result of the formula t = CBS/CIR provides the guaranteed number of bits over a period of time (t) that a particular service will transfer. For example, a CIR of 256 Kbps and a CBS of 512 kilobits (Kb) means that the network will move 512 Kb in any given 2-second period (2s = 512 Kb/256 Kbps); this is the guaranteed rate for periods of congestion.

Under light loads, the network's actual throughput will be greater than the CBS; this is referred to as excess burst size (EBS). Other important terms in a frame relay environment are:

- Frame relay access device (FRAD) provides access to a frame relay network. In a LAN, a FRAD is typically a router.

- UNI specifies the signaling and management functions between a frame relay network device and end user's device.

- Network-to-Network Interface (NNI) specifies the signaling and management functions between two frame relay networks.

Permanent and Virtual Circuits

In frame relay networks, applications share available bandwidth. However, frame relay can give active applications full access to the network's bandwidth when no other applications are sending or receiving information. Because frame relay switches have the intelligence and capability to obtain and interpret network status, and take corrective action, frame relay can reroute traffic when problems occur. Typical private lines do not have this capability.

Frame relay carries network traffic over virtual circuits. These circuits are mapped from end-to-end through one or more NNIs. More than one virtual circuit may be mapped over an NNI, and each virtual circuit takes its turn accessing the bandwidth available at the NNI's ports.

The amount of time each virtual circuit has on the port depends on the circuit's CIR. The telecommunications carrier assigns each circuit a CIR as it provisions (creates) the circuit. Depending on the circuit provisioning, a circuit can take advantage of periods of low network usage across the NNI by bursting above its CIR.

To illustrate how a frame relay circuit can adjust to increased bandwidth requirements, compare TDM to frame relay as shown on the Frame Relay Bursting and Performance Diagram. In TDM, a larger file is given more time to move over the network. In frame relay, a larger file is given more bandwidth over the network, using the same period of time as a smaller file. As above, this depends on how the carrier provisions the circuits.

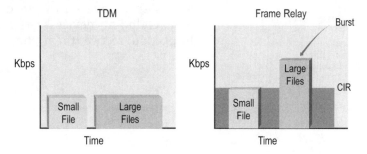

Frame Relay Bursting and Performance

Frame Relay Operation

The frame relay protocol is straightforward; frame relay network devices are only responsible for moving frames from the source to the destination. If invalid frames are received by a frame relay device, they are discarded without sending notification to a sending or receiving station. No sequencing of frames is supported, and no control information is sent in the frame information field, only user data. Frame relay devices do not acknowledge frames at the receiving end of the network.

Frame Relay Frame

A frame relay frame is based on the HDLC Link Access Procedure for D Channel (LAPD) frame format. It consists of the following fields, as shown on the Frame Relay Frame Format Diagram:

- **Flag**—A flag, 7E, indicates the start and end of each frame.

- **Frame relay header field** (discussed below).

- **Variable-length information field**—This field contains user data. The maximum recommended payload is 1,600 bytes; the minimum payload is 1 byte. Frame relay devices in a network disregard the content in the information portion of the frame.

- **Frame check sequence (FCS) field**—The FCS, which is locally significant only, is used to check that the frame has been received without errors. At the end of each frame, an FCS is submitted by the access device to ensure bit integrity. Frames that have errors are discarded. Unlike X.25, frame relay endpoint devices recognize that frames have been dropped; to recover them, they request transmission.

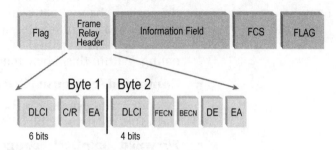

DLCI: Data Link Connection Identifier
C/R: Command/Response Field Bit
EA: Address Extension Bit (allows 3- or 4-byte header)
FECN: Forward Explicit Congestion Notification
BECN: Backward Explicit Congestion Notification
DE: Discard Eligibility Indicator

Frame Relay Frame Format

Information Field

LAPD's information field (I field) contains data passed between devices over a frame relay network. The I field is variable in length. Although the theoretical maximum integrity of an FCS is 4,096 bytes, the actual maximum integrity is vendor-specific. Frame relay standards ensure the "minimum maximum" value supported by all networks is 1,600 bytes.

User data may contain various types of protocols (PDUs), which are used by access devices. According to Internet Engineering Task Force (IETF) Request for Comment (RFC) 1490, an industry-standard mechanism for specifying which protocol is in the I field, the I field may also include "multiprotocol encapsulation." With or without multiprotocol encapsulation, protocol information sent in the I field is transparent to a frame relay network.

Frame Relay Header Field

The frame relay header field is key to the operation of a frame relay virtual circuit. A frame relay header consists of:

• **Data Link Connection Identifier (DLCI)**—The DLCI identifies the virtual circuit for a frame. These 10 bits allow for over one thousand virtual circuit addresses for each NNI physical interface. The remaining header bits are used for congestion and circuit control, and are listed on the Frame Relay Frame Format Diagram. Frame relay devices derive this 10-bit address from the first 6 bits of the first byte, and the first 4 bits from the second byte of the frame relay header. The DLCI identifies the logical channel between a user and network, and has no network-wide significance. Supervisory (control) information passed by means of the DLCI is considered "out-of-band" signaling. Within this 2-byte frame relay header is the 10-bit data.

- **Command/response (C/R) bit**—The C/R bit is not used.

- Extended address (EA) bits—EA bits are used to extend an address from 10 to 12 bits. EA bits are typically not used.

- **Forward explicit congestion notification (FECN) bit**—The FECN bit is set to notify a user that congestion was experienced as the frame traversed the network. It is set by the network, not the user.

- **Backward explicit congestion notification (BECN) bit**—The BECN bit is set to indicate congestion may be experienced by traffic sent in the opposite direction.

- **Discard eligibility (DE) bit**—Data terminal equipment (DTE) on a frame relay network, such as a router, may set the DE bit to indicate the frame is less important than other frames without the DE bit set. To manage congestion and fairness, other frame relay devices may discard this frame.

Flow Control

Frame relay specifications provide a method for flow control; however, they do not guarantee implementation of those standards on devices. This is a vendor-specific issue that is often a key difference in the performance of vendors' products; however, it does not generally interfere with basic frame relay interoperability.

Congestion

Frame relay uses bits in the header to indicate network congestion. A network may send congestion condition notifiers to access devices through FECN and BECN bits. Access devices are responsible for restricting data flow under such congested conditions.

The Frame Relay BECN/FECN Diagram shows how messages would be transmitted if congestion was occurring at Device C. Device C detects congestion in the direction of Device E. Device C sets the FECN bit on frames destined for Device E; Device E knows to redirect or drop frames as necessary to avoid congestion. Device C also sets the BECN bit on frames destined for Device A, informing it of congestion between Devices C and E. Device A takes necessary action to avoid congestion between Devices C and E.

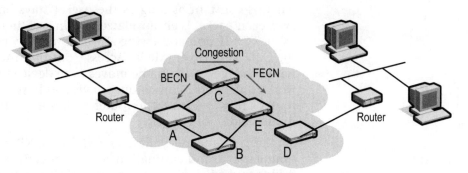

Frame Relay BECN/FECN

Management Frames

Network and access devices may pass special management frames with unique DLCI addresses. These frames monitor status link, and reflect whether it is active or inactive. Management frames also pass information regarding the current status of permanent virtual circuits (PVCs), as well as any DLCI changes within the network. Local Management Interface (LMI) is the protocol used to provide information about PVC status. Frame relay's original specification did not provide for this kind of status. Since then, American National Standards Institute (ANSI) and Consultative Committee for International Telegraphy and Telephony (CCITT) specifications developed and incorporated a method for LMI, now known "officially" as Data Link Control Management Interface (DLCMI).

Frame Relay Addressing

A frame relay connection is a type of virtual circuit, referred to as a data link connection (DLC). In the majority of frame relay equipment, and therefore services, DLCs are PVCs, predefined by both sides of the connection. SVCs are also defined in frame relay specifications and are gaining in use. The original service offerings for frame relay were PVC-based, a trend that has continued to the present day. PVCs efficiently serve the needs of most existing data applications. There is, however, growing vendor support and implementation for SVC capabilities, to meet emerging applications and stimulate intercorporate communications.

Each DLC, whether a PVC or SVC, has an identifying DLCI. A DLCI changes as a frame travels from segment to segment, similar to the physical address changes that occur in a routed Ethernet LAN. In an Ethernet LAN, each node has a network address. The destination network address is included in each packet, and remains constant as long as the packet stays on the LAN. However, each packet is encapsulated within an Ethernet frame. When using TCP/IP and addressing a packet to a device across a router, the Data Link Layer (Media Access Control [MAC]) address of the frame changes as the frame moves from device to device, and network to network. The sending device addresses the frame to the router's MAC address, the router addresses its frame to the next router, and so on.

Similarly, frame relay devices (NNI or UNI) change the DLCI (frame source and destination addresses) as the frame moves from device to device. Thus, we say the DLCI has "local significance," because each DLCI only pertains to one local segment of the frame relay network. The initial DLCI, that the sending FRAD uses to address the frame to the next frame relay device, may not be the same DLCI that the final frame relay device uses to pass the frame to the FRAD at the destination network.

Therefore, we need to know both the local and remote DLCIs for any link we are provisioning across a frame relay network. Routing tables in each intervening frame relay switch, whether in the carrier's network or a private network, take care of directing frames to the proper destination, alternately reading and assigning DLCI values in the control portion of the frames, as appropriate.

A DLCI is the portion of the frame that identifies the logical channel between an end device and network. Because the DLCI only identifies the connection to the network, it is up to the network devices to map the two communicating DLCIs. The Frame Relay Addressing Diagram illustrates this scenario.

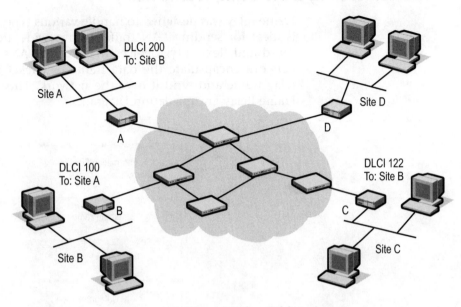

Frame Relay Addressing

In this diagram, DLCI 200 is routed by the frame relay network from Site A to Site B, where each device alters the DLCI to match the local link. The frame finally appears at Router B as DLCI 100. All data sent from Site B to Site A is addressed to DLCI 100, and appears as a DLCI 200 at Site A. Improper configuration of DLCIs is a common mistake. Frame relay is designed to carry both Data Link Layer frames from other topologies as well as Network Layer packets.

Interconnecting LANs Using Frame Relay

Frame relay was designed to handle various types of LAN traffic. It is ideal for sending LAN traffic over a wide area because of its speed and flexibility. Devices that attach LANs to a frame relay network encapsulate the data frame or packet inside the frame relay frame and send it over the network. This is shown on the Frame Relay Encapsulation Diagram.

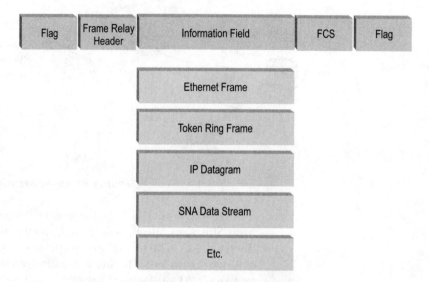

Flag	Frame Relay Header	Information Field	FCS	Flag

| Ethernet Frame |
| Token Ring Frame |
| IP Datagram |
| SNA Data Stream |
| Etc. |

Frame Relay Encapsulation

At the other end of the network, information is removed from the frame relay frame and sent to the final destination.

Frame Relay Implementation Options

Frame relay, as with many technologies discussed in this course, may be implemented in several forms. Conceptually, there are three ways to implement a frame relay network, including:

- Using private network facilities
- Using public facilities
- Incorporating a hybrid solution, using both private and public facilities

Most organizations are adopting neither a pure public nor pure private network architecture, but a combination of the two: a hybrid network design. This has become a standard architecture in the industry among large organizations. That is why it is important to evaluate public-switched telephone network services side-by-side with private line services when designing and implementing a WAN.

Private Frame Relay

A private frame relay network is constructed using transmission links and frame relay equipment owned and operated by a customer. In this example, illustrated on the Private Frame Relay Network Diagram, a private frame relay network connects several of an organization's sites.

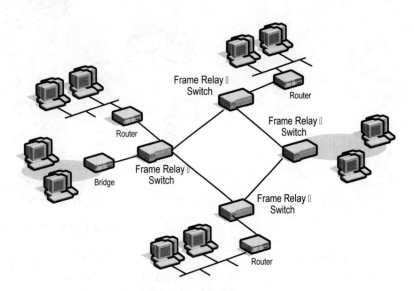

Private Frame Relay Network

**Public Frame
Relay**

A frame relay network can also be implemented through the use of public facilities. In this case, the frame relay switches and backbone are owned and operated by a telecommunications service provider. A customer does not have insight into the switches and switch configurations or paths the information takes from source network to destination network. The customer does not have to manage the network either. The cloud on the Public Frame Relay Network Diagram illustrates this concept. Virtual circuits are provisioned for each specific site and can be used only by that end user. This type of network is called a virtual private network (VPN).

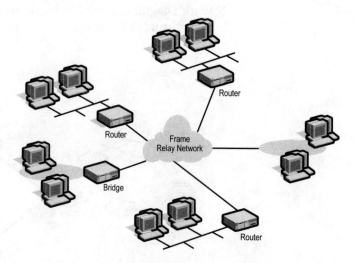

Public Frame Relay Network

**Hybrid Frame
Relay**

Private and public facilities and equipment can be combined to form a hybrid frame relay network. The decision to use public, private, or a mixture of the two is made based on each business application. It may also be necessary to combine public and private facilities or other technologies if service is not available in a particular area that needs connectivity to the overall network. The Hybrid Frame Relay Network Diagram illustrates the concept of a hybrid frame relay network.

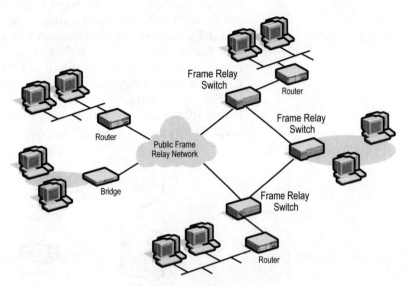

Hybrid Frame Relay Network

End-User Access to Frame Relay Networks

End-user devices can access frame relay networks in a wide variety of ways. Typically, PCs and workstations use frame relay networks, as do many voice and video applications. End-user devices include:

- PCs
- Workstations
- Controllers
- PBX equipment

End-user devices are connected to customer premises equipment (CPE), such as bridges and routers, that have frame relay capabilities. CPE devices take information from a network and place it into frame relay frames. The frame relay frames are passed to the frame relay switch by means of a UNI connection between the CPE and frame relay network. There are many connection possibilities, as illustrated on the End-User Connectivity Diagram. Frame relay switches within the frame relay network provide the connectivity between endpoints over NNI.

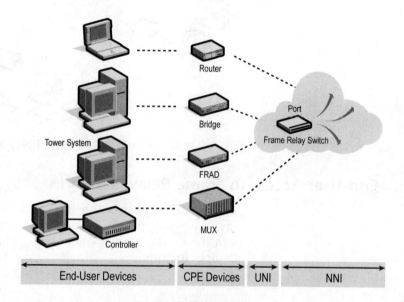

End-User Connectivity

The "Extra Hop" Problem

As discussed earlier, frame relay can transport a wide range of Network Layer protocols. However, a key assumption of TCP/IP networking can reduce the efficiency of a frame relay network.

Classical IP routing requires that each subnet communicate with other subnets through a designated default router. If a subnet's default router is located across the frame relay cloud, transmissions from that subnet must travel over additional hops, slowing overall network performance.

To see how this works, refer to the Extra Hop Problem Diagram. Let us say that a host on Subnet A has an IP transmission for a host on Subnet C. According to the rules of TCP/IP networking, any host on Subnet A must forward its transmission first to its default Router A, which will then forward it to the routers that serve other subnets.

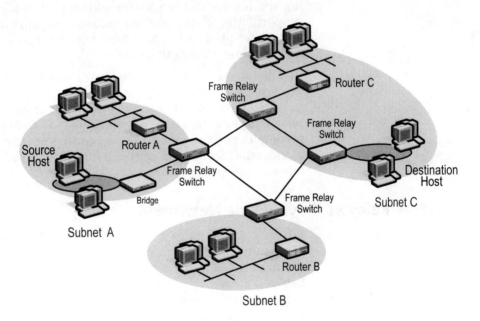

Extra Hop Problem

The "extra hop" problem occurs because the source host and Router A are separated by part of the frame relay network, as are Router C and the destination host. To follow the classical IP rules, the transmission must first travel from the source host to Router A, through a frame relay switch. Next, the frame relay network sets up an SVC from Router A to Router C. Then Router C forwards the transmission to the destination host, back through the frame relay cloud.

This is a very inefficient way to set up the virtual circuit, because both the source and destination hosts have more direct access to the frame relay network. Many hops could be saved if they could leave the routers out of the transmission, and simply set up the most direct virtual circuit straight across the frame relay cloud. A proposed Internet standard, Non-Broadcast, Multi-Access (NBMA) Next Hop Resolution Protocol (NHRP), is intended to do this, bypassing the classical IP rules that require routers to control IP transmission.

The operation of NHRP is similar to that of Address Resolution Protocol (ARP). In the frame relay network, certain switches are designated as next hop servers (NHSs). Each NHS gradually learns the locations of destination hosts, by monitoring traffic and sending specific address request messages. If all frame relay switches support NHRP, they can then use route information cached in an NHS to set up direct connections between hosts, without relying on each subnet's default router.

NHRP is an emerging technology, still under development by IETF. While it solves the extra hop problem, NHRP currently has the potential to create other routing problems of its own. However, as private and public cloud networks such as frame relay and ATM become larger, the extra hop problem becomes more of a performance issue. Some form of NHRP is likely to be adopted to streamline transmission across these wide area switched networks.

Frame Relay vs. Private Line Networking

Private point-to-point lines (such as T1) are widely used in computer networks to connect sites. Private lines use TDM for communication of traffic across a WAN. This section differentiates between a native point-to-point T1 service, and a frame relay service that uses various Physical Layer services, including T1.

The emphasis in wide area networking is shifting away from dedicated networks toward switched alternatives. The reason is because the higher quality of the public-switched network makes switched alternatives highly reliable and efficient. The latest generation of public network services, including frame relay and ATM services, is continuing this trend.

One of the reasons there is a growing demand for frame relay is the decrease in complexity of a frame relay network vs. a T1 network. As the number of endpoints grows, frame relay is a simpler way to provide all-to-all connectivity. This is illustrated on the T1 vs. Frame Relay Network Diagram.

Frame Relay Network

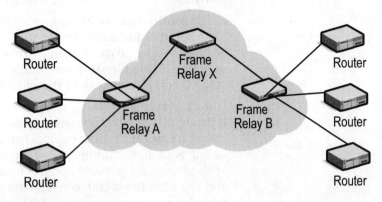

T1 Mesh Network

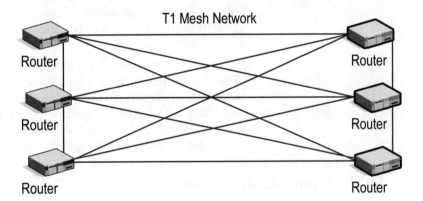

T1 vs. Frame Relay Network

Typical frame relay implementations use T1 as the Physical Layer service. In other words, frame relay rides on top of a T1-based point-to-point service, because frame relay switches are connected by T1 lines. However, when a customer must interconnect many sites, frame relay is a more efficient way to provide T1 access speeds between those sites.

Frame Relay Services

Today, frame relay service can deliver data at very high rates. Source and destination points communicate from their locations to the frame relay cloud over leased-line connections, which generally are fractional or full T1 connections.

To transmit data to its proper destination, frame relay frames incorporate addressing information, which the network uses to ensure proper data routing through the service provider's switches. This addressing essentially lets users establish virtual circuits to communicate over the same access links.

Part of what makes frame relay so attractive to network managers is its use of a public data network, or cloud, that enables an organization to minimize line, equipment, and management costs associated with maintaining its own complex mesh-topology WAN. Such savings are possible because most of the burden of designing and maintaining a frame relay data network falls on the service provider, companies such as AT&T, MCI, and Qwest.

There are many advantages to this arrangement. For one, it relieves a network manager of the burden of managing the overall infrastructure. It also reduces the amount of effort necessary to make needed changes.

If a network manager needs to increase the bandwidth or number of access points to the cloud, for example, he or she will most likely only need to make a few telephone calls to the provider, and then make slight adjustments to the central-site router. If additional speed is needed for new applications, in many cases, added bandwidth can be provided on demand.

Public Frame Relay Services

The first public frame relay service was introduced in the United States in 1991. There are currently a large number of service providers in the frame relay market. Services provided are illustrated on the Frame Relay Public Services Diagram, and include:

- Basic frame relay transmission

- Multiple access alternatives

- Customer network management

- Internet access

- International connectivity
- Managed network services
- Frame relay to ATM internetworking

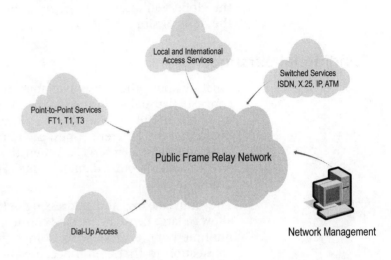

Frame Relay Public Services

As we can see on the diagram, there are many ways to access a frame relay network. Analog dial-up access using modems is a suitable choice for mobile workers, telecommuters, and occasional network users. Switched access options, such as ISDN and Switched 56 (SW56), are also available.

Bandwidth on Demand

Frame relay is well suited to bursty data traffic, because it can provide bandwidth on demand. It does this by using statistical multiplexing, which does not require that a link be up and dedicated at all times. Rather, frame relay allocates bandwidth only when data is ready to be transmitted.

Other, more traditional WAN transports use TDM, whereby each data transmission requires dedicated bandwidth across a WAN. A disadvantage of this method is that even "data silence," periods when the network is passing no data, requires the link to be up and dedicated.

Unlike a leased line, frame relay can be set up with two connection speeds: CIR and excess information rate (EIR). CIR is the guaranteed average bandwidth available, determined by our estimate of normal traffic. If network traffic increases past the CIR, the cloud will attempt to open additional circuits and complete the transmission.

Conditional Bursting

Bursting above the CIR is available only when the network is not congested, usually during non-peak periods. This capability is especially useful for branch offices located in different time zones. Because of the different zones, each branch office would reach its peak and burst at intervals. When the network is not congested, we can actually burst data, in some cases, at capacities up to two times the CIR.

The multiplexing and addressing scheme used by frame relay allows a large central site to be connected to the cloud (and hence multiple remote sites), with a single router port and high-speed connection to that cloud. Because the circuits are not dedicated on a conversation-by-conversation basis to any specific remote site, many remote-to-central site transmissions can take place simultaneously.

VoIP over Public Frame Relay Services

The nature of frame relay poses quite a challenge for VoIP across frame relay links. Recalling the details of frame relay from earlier in this lesson, the most common virtual circuit is a PVC, which is associated with a CIR. The access circuit rate and the frame relay port rate must match. The CIR of each PVC is sized so that it is half the respective port rate, which is a common implementation. Each branch office is guaranteed its respective CIR, but it is also allowed to burst up to the port rate without any guarantees. This can cause latency issues when interLATA transport is involved.

The obstacle in running VoIP over frame relay involves the treatment of traffic within the CIR and outside of the CIR, commonly termed the "burst range." Traffic up to the CIR is guaranteed, whereas traffic beyond the CIR is not. This is how frame relay is intended to work. CIR is a committed and reliable rate, but burst is a bonus when network conditions permit it without infringing upon any other user's CIR. For this reason, burst frames are always marked as Discard Eligible (DE) and are queued or discarded when any network congestion exists. Most customers can achieve significant burst throughput, but it is unreliable, unpredictable, and not suitable for real-time applications like VoIP.

By sizing the CIR to meet or exceed the peak voice traffic, and then applying priority queuing on the interface so that VoIP is serviced first, we can intuitively assure that voice traffic will not enter the burst range. Actual queuing mechanisms, however, are not always intuitive. Even though the aggregate voice traffic cannot exceed the CIR, it is still possible that a voice packet could be sent in the burst range.

The good news is that most interexchange carriers (IXCs) convert the long-haul delivery of frame relay into ATM right after leaving and right before entering the customer's premises. ATM has a built in class of service (CoS) and CBR. ATM cells are delivered with lower latency and higher reliability. It should be understood, however, that even under the best circumstances, frame relay is still more susceptible to delay than ATM or TDM.

Cost Considerations

If we have multiple sites in several geographically distant locations, or our organization plans to add multiple sites across the country in a relatively short period of time, frame relay may save us from putting a significant crimp in our telecommunications budget.

Frame relay is a very cost-effective technology to transport data between local access and transport areas (LATAs) (interLATA transport). Frame relay may not be financially attractive (or even possible) for multiple sites in the same LATA, that is, several sites in the same metropolitan area.

Key to determining the cost effectiveness of frame relay is the expense of running a 56-Kbps or T1 (1.544-Mbps) dedicated leased circuit from each office to a frame relay service provider. If the number of sites is low, and all sites are in the same area, it may be cheaper to run dedicated lines to each of them, rather than use frame relay.

Carrier Choice Considerations

Although frame relay is a relatively mature technology, there are still some moving targets an organization may want to clarify before executing an agreement with a specific carrier.

CIR

When we subscribe to a frame relay service, we select a Committed Information Rate (CIR) for our network connection. The CIR defines the average bandwidth provided by the carrier over a given period of time. We select the CIR, depending on the carrier, to match the expected performance across our WAN connection. Typical CIRs range from 56 Kbps to 1.5 Mbps.

Unfortunately, the selected CIR is not always guaranteed. The business success of carriers hinges on building networks only as demand requires. A general rule of thumb seems to be that a carrier will add links to its network when the network becomes 70 to 80 percent subscribed. Often, congestion is really just a delicate term for an oversubscribed network.

Congestion Control

Thus, with any prospective frame relay provider, we must explore the critical issue of how its network handles "congestion control," which differs widely among carriers. In a nutshell, when some networks get too congested, they discard data frames. Because frame relay performs little to no error detection and correction, the user's network is responsible for detecting discarded frames and requesting retransmission.

Reports and Statistics

It is also good to find out what management reports and systems the carrier provides, and whether up-to-the-minute statistics on utilization percentages, traffic patterns, and frame discard/error rates are available.

Frame Relay Network Requirements

After we have selected and signed on with a carrier, we must purchase access to the carrier's frame relay network. This means that to use the frame relay network, we need to connect an appropriately high-speed pipe from our organization's site to the frame relay network.

The hardware and software configuration required to connect to a frame relay network is not complex. Each site needs a router with a WAN-connection port (or ports) appropriate to its network protocol, software/firmware for the router that supports frame relay, and a DSU/CSU network interface.

Configuration of the router involves the simple input of the appropriate DLCI information furnished by the frame relay service provider in the configuration tables. Assuming all the various connections (LAN, router-to-telephone company, and telephone company-to-frame relay network) are physically correct and activated, the frame relay network connection should be established.

Activities

1. Research frame relay vs. X.25 services. How does frame relay differ from X.25 networks? What benefits does X.25 provide over frame relay? What benefits does frame relay provide over X.25?

   _____

   _____

   _____

   _____

   _____

   _____

2. Frame relay is well-suited to voice and video applications. True or False?

3. CIR is the actual frame relay network throughput realized at times of low usage. True or False?

4. CBS defines the number of bits that can be transferred during some time interval. True or False?

5. A device that provides access to a frame relay network is a:

   a. ECS

   b. FRAD

   c. NNI

   d. UNI

6. What protocol is most susceptible to delay?

   a. Asynchronous Transfer Mode

   b. Frame relay

   c. Time division multiplexing

   d. They all have similar delay susceptibility.

7. CIR is the:

    a. Guaranteed average data rate for a service

    b. Number of bits that can be transferred over a period of time

    c. Throughput that can be realized during periods of low network usage

    d. Signaling and management functions between two frame relay devices

8. Draw a protocol stack and corresponding frame relay frame that encapsulates a TCP/IP packet.

9. Five locations are geographically dispersed across the country. Routers are in place at each location to provide LAN interconnection service across the country. Each router is connected to many LANs within a region, such as different buildings across a city. We will assume the traffic is intermittent, but has high peak rates. Users have point-to-point leased lines of different capacities across the nation, dependent upon the traffic pattern.

The non-frame relay network, as depicted on the Initial Network Diagram, consists of 5 routers and 14 DSU/CSUs, 4 leased lines rated at 56 Kbps, 2 leased lines rated at 1.544 Mbps (T1), and 1 leased line rated at 256 Kbps (FT1). There are 14 router ports which are incremental costs to the routers. In addition, most router manufacturers tie router performance to the amount of memory purchased with the router.

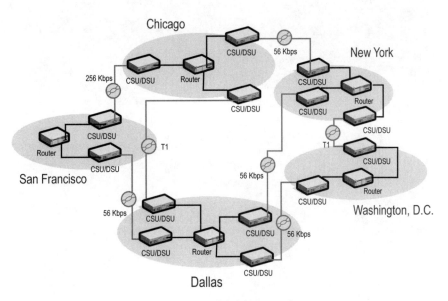

Initial Network

If you look closely at the diagram, you should note the following:

- Seven leased lines are being used of varying capacity and distance.

- Fourteen DSU/CSUs are being used.

- Fourteen router ports are being used.

Determine what is needed for a frame relay implementation. Draw a diagram that will meet the above needs using frame relay.

Extended Activities

1. Locate a white paper on frame relay from a networking hardware vendor. Write a summary of the paper.

   _____

   _____

   _____

   _____

   _____

   _____

   _____

   _____

   _____

   _____

   _____

   _____

   _____

   _____

   _____

   _____

   _____

   _____

   _____

   _____

   _____

   _____

2. Go to the Frame Relay Forum Web site **http://www.frforum.com** and review the latest information on frame relay. Review the frame relay overview provided by the Frame Relay Forum.

3. Review the following Web sites and note the information provided on frame relay. List the products that have frame relay support, and state what each would be used for in wide area networking.

   a. Nortel   **www.nortel.com**

   b. Adtran   **www.adtran.com**

   c. 3Com   **www.3com.com**

   d. Discount Datacomm   **www.discountdata.com**

   e. Avaya   **www.avaya.com**

Lesson 6—Switched Multimegabit Data Service

Switched Multimegabit Data Service (SMDS) is a MAN and WAN service designed to provide bandwidth on demand. It was originally designed to provide connections between LANs in a city-wide region and within a single LATA. SMDS was developed by Bellcore (now Telcordia, formerly an AT&T subsidiary). Several Bell operating companies have made SMDS available on their networks.

Objectives

At the end of this lesson you will be able to:

- Show how the SMDS protocol stack relates to the OSI model

- Describe the type of traffic SMDS is best suited to transport, and explain why

- Explain how traffic passes from a LAN, through a router and DSU, and into an SMDS network

 **Key Point**

> *SMDS is a connectionless, cell-switched service that delivers data rates up to 45 Mbps.*

SMDS Operation

The basic characteristics of SMDS are largely described by its name.

Switched

SMDS is a public-switched network, similar to frame relay. To the networks that connect to it, SMDS has no "distance." In other words, like the public telephone system, all that is required to send a packet to another SMDS-connected network is the user's address. To a subscriber's LAN, SMDS looks like another subnet. All internal nodes of the SMDS "cloud" are hidden from the subscriber.

SMDS is a connectionless, best-effort service; it is not necessary to set up an end-to-end connection before sending data. SMDS is also a cell-relay system. Like ATM, it uses 53-byte cells that each carry a 48-byte payload; however, SMDS uses a different format for its 5-byte cell header.

Multimegabit

SMDS falls in the same performance range as frame relay (T1/E1 to T3/E3), potentially overlapping the low end of B-ISDN. It can provide up to 45 Mbps of bandwidth, because each pair of SMDS switches is connected by two one-way T3 lines (typically fiber optic cable).

SNI

The SMDS protocol specifies how to connect CPE with an SMDS network. The point at which the CPE interfaces with the SMDS network is called the Subscriber Network Interface (SNI). This interface, connecting a customer to the SMDS cloud, is usually implemented by means of T1 or T3 lines, as shown on the SMDS Network Example Diagram.

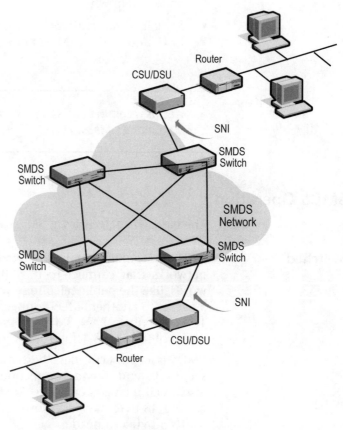

SMDS Network Example

SMDS Protocols

SMDS protocols were defined by the regional Bell operating companies (RBOCs), based on the Institute of Electrical and Electronic Engineers (IEEE) 802.6 standard for MANs. This standard was originally known as Queued Packet and Synchronous Exchange (QPSX), and is now called Distributed Queue Dual Bus (DQDB). Each DQDB end station has two buses; one for incoming traffic and one for outgoing traffic. Thus, DQDB provides full-duplex communication between any two nodes. However, an SMDS network supports only connectionless data traffic.

The interface with SMDS is the Logical Link Control (LLC) sublayer of the OSI Data Link Layer, as shown on the SMDS Layers Diagram. Thus, while SMDS uses cell relay internally, it is technically a frame-switching facility because LLC deals with frames. You are not likely to hear it called that, however.

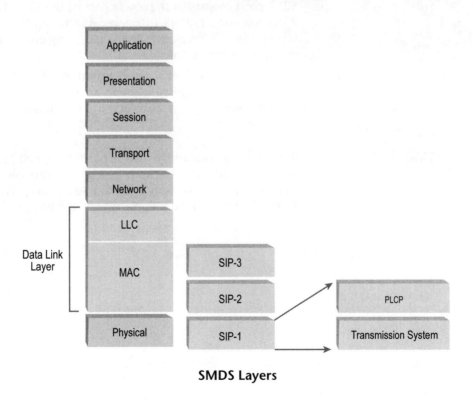

SMDS Layers

SMDS Interface Protocol

The SMDS protocol stack is known as the SMDS Interface Protocol (SIP). This stack consists of internal SMDS protocols called SIP-1, SIP-2, and SIP-3.

SIP-1

SIP-1 corresponds to the OSI Physical Layer. SMDS uses cell relay at Layer 1, relaying 53-byte cells similar to ATM; however, the format of the 5-byte header is different.

The SIP-1 layer is divided into two sublayers:

- Physical Layer Convergence Protocol (PLCP) sublayer defines how the SMDS 53-byte cell is mapped to the specific transmission system described at the lower sublayer.

- Transmission System sublayer describes the digital carrier used for SNI. This includes DS1 (T1), DS3 (T3), SONET, and High Speed Serial Interface (HSSI).

SIP-2

SIP-2 corresponds to the lower part of the OSI MAC sublayer. SIP-2 defines the network interface. At the sending end, it segments frames into 53-byte cells. At the receiving end, it reassembles individual cells into frames.

SIP-2 uses Queued Arbitrated functionality, which allocates cells to each data transmission on an as-needed basis. This approach is most appropriate for connectionless transmissions that are not time sensitive.

SIP-3

SIP-3 corresponds to the upper part of the OSI MAC sublayer. SIP-3 transports data from the upper layer protocols. It is also responsible for error correction and data addressing. The SIP-3 data unit is the PDU (described below), which can be as large as 9,188 bytes.

DXI

The SMDS Interest Group (SIG) created the Data Exchange Interface (DXI) protocol to provide a standard interface between a router and the DSU used to access an SMDS network. DXI only exists to move data between a router and DSU/CSU; it is not used in every network. The SMDS and Data Exchange Interface Diagram illustrates this relationship.

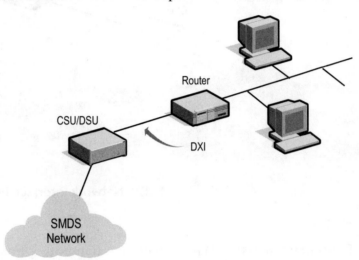

SMDS and Data Exchange Interface

The DXI specifies how various SIP-layer tasks are distributed between the router and DSU. The SMDS Network Interface Points Diagram presents the protocols associated w ,ith the router, and protocols associated with the DSU. As we can see on the diagram, DXI is divided into the DXI Link and DXI Physical Layers.

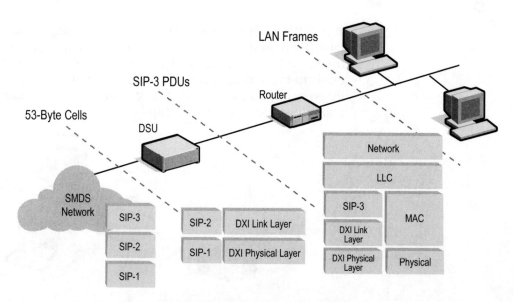

SMDS Network Interface Points

Router Functions When frames are transmitted from a LAN to a router, a router converts the MAC layer protocol (such as Ethernet) to a SIP-3 PDU for delivery to the DSU. The DXI Link Layer encapsulates each PDU between a header and trailer, creating a frame. The purpose of this frame is to move PDUs across the physical link between the router and DSU. DXI Link Layer protocols are based on the HDLC protocol, thus the format of the frame is very similar to HDLC.

The DXI Physical Layer defines a connection between a router and DSU. The most common DXI Physical Layer protocols are HSSI, V.35, and X.21.

DSU Functions A DSU receives the DXI Link Layer frames. It removes SIP-3 PDUs from frames, then converts them into cells for transmission across the SNI to the SMDS switch.

The opposite happens when cells are received. The DSU converts cells to PDUs, then encapsulates them into frames. The frames are transmitted across the DXI to the router. The router removes PDUs from frames, then converts PDUs back into frames for the LAN. This process is summarized on the SMDS Encapsulation Diagram.

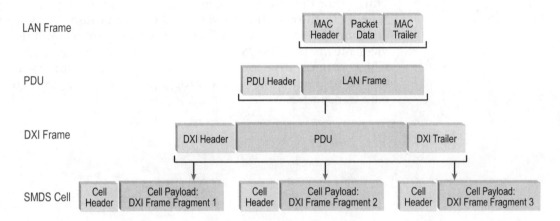

SMDS Encapsulation

Advantages and Disadvantages

Like frame relay, SMDS is ideal for data transfers, and city-wide client-server computer networks with many-to-many communications. In other words, if many locations must exchange data with one another (not just back to the home office), SMDS can be an effective solution. It is especially cost-effective if the locations must transfer data at data rates higher than T1.

Although SMDS scales down, most carriers only offer it at up to 45 Mbps. This makes it useful for businesses needing bandwidth up to DS3 speeds, but less desirable when ATM speeds are required. (ATM is a better choice in this case.) Note that the nodes connected by an SMDS network need not operate at the same speed; one site's SNI may operate at a lower or higher bandwidth than another. Additionally, since SMDS is considered a "bursty" data service, each node can handle occasional increases in traffic without a subsequent increase in connectivity costs.

SMDS is not a good choice for WANs that cross LATA boundaries, because of the restrictions placed on LECs in the 1984 Modified Final Judgment (MFJ). This rule has limited SMDS to a metropolitan area service only, which is not attractive to business customers who want to exchange data with locations outside their LATAs. To set up this sort of interLATA network with SMDS, a LEC must partner with an interexchange carrier (IXC).

Even though SMDS is a cell-based technology, it does not perform very well for interactive real-time video or voice, or pay-per-view movies. This is because it is also a connectionless technology. Thus, while data travels in same-size cells, the cells may take multiple paths through the network, and that can cause delays at the destination end. The best-effort approach of connectionless services means that SMDS cannot guarantee delivery of each cell, which is imperative for voice and video communications. Thus, despite its potential for faster throughput, SMDS is best suited for data transport.

Activity

Given the following diagram, describe the movement of information from Client A to Client B and back. In your discussion, describe what protocols are used at each device and across the WAN.

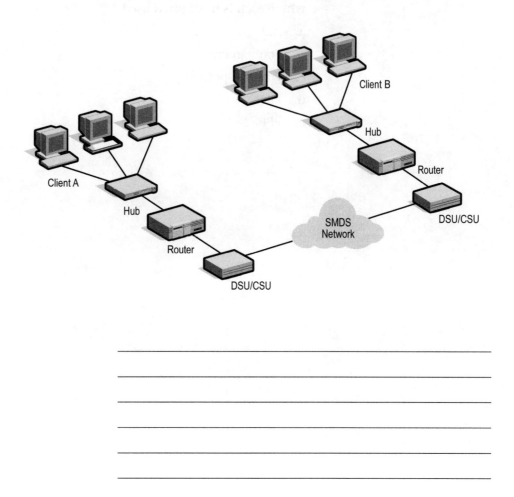

_____

_____

_____

_____

_____

_____

Extended Activity

1. Compare the protocols below to SMDS. Describe the differences and similarities to SMDS based on the way connections are established, the size and type of frame or packet, and where each is most often used.

    a. X.25

    b. Frame relay

    c. Point-to-point T1

    d. ISDN

    e. PPP

Lesson 7—X.25

X.25 was the first connectionless Network Layer protocol. It is still commonly used to switch packet traffic over a wide area.

An X.25 network, whether public or private, is typically built largely upon the leased-line facilities of the public telephone network. It uses a Network Layer address (telephone number) so that switches can route traffic by means of multiple paths.

In the US, although an installed base exists, new X.25 services have all but been replaced by faster technologies such as frame relay and IP VPNs. Nonetheless, understanding the X.25 protocol and services will help you understand faster and more efficient protocols, such as frame relay and ISDN, which were built upon the foundation of X.25.

Objectives

At the end of this lesson you will be able to:

- Describe how X.25 is used to transport data over a wide area

- Explain the difference between packet switching, frame switching, frame relay, and cell relay

- Name the X.25 protocol layers and describe their functions

- Explain what a packet assembler/disassembler (PAD) is used for

 **Key Point**

Today's fast packet-switched networks are based on X.25.

X.25 Services

X.25 is connection-oriented, and offers two types of service:

- **PVCs**—This is the X.25 equivalent of a leased line, statically defined and always available as long as a network is up. Unlike leased lines, however, more than one virtual circuit can share a physical link.

- **Virtual Connections**—This is the X.25 equivalent of a dial-up connection. A network establishes a connection on a virtual circuit, transfers packets until the application is finished, and then releases the connection.

Addressing

ITU-T recommendation X.121 defines a system of assigning addresses to devices on an X.25 packet network. The X.121 system is similar to the numbering plan used for voice telephone networks. Every X.25 user, anywhere in the world, is uniquely identified by a Network Layer address that includes codes for world zone, country, network, and individual user. Thus, an X.25 communication can take place between any two X.25 users, as long as they are both on the same network, or interconnected networks.

X.25 Protocols

The X.25 interface lies at OSI Layer 3, because X.25 provides error-free service to the Transport Layer. However, as we will see later, the overhead associated with the necessary error checking is proving to be unacceptable in today's highly reliable digital networks.

X.25 defines its own three-layer protocol stack, as illustrated on the X.25 Protocol Layers Diagram. The X.25 standard predates OSI (the first version of X.25 was issued in 1976). OSI has adopted X.25 Layer 3 as a connection-oriented Network Layer protocol.

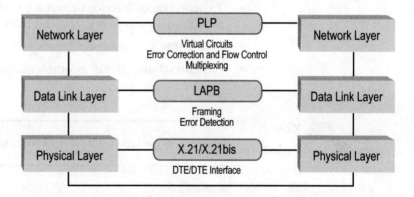

X.25 Protocol Layers

The X.25 standard does not itself fully define all three layers of the stack, but rather refers to other standards. X.75, for example, is a standard that defines the interface between two distinct X.25 networks and is nearly identical to X.25. X.25 consists of these protocols:

- **Layer 3**—Packet Layer Protocol (PLP)
- **Layer 2**—Link Access Procedure-Balanced (LAPB)
- **Layer 1**—X.21 and X.21bis

Layer 3: PLP Packet Layer Protocol (PLP) operates at the OSI Network Layer. PLP manages connections between data communications equipment (DCE) and DTE anywhere in a network. It accepts data from a Transport Layer process, breaks the data into packets, assigns the packets a Network Layer address, and takes responsibility for error-free delivery of the packets to their destination. PLP establishes virtual circuits and routes packets across the circuits. Because many virtual circuits can share a link, PLP also handles multiplexing of packets.

Layer 2: LAPB Link Access Procedure-Balanced (LAPB) operates at OSI Layer 2, and provides full-duplex point-to-point delivery of error-free frames across a link. These frames deliver packets to and from processes operating at Layer 3.

LAPB is a subset of the ISO HDLC standard. It is "balanced" because the LAPB standard excludes portions of the HDLC standard having to do with multidrop, "unbalanced" operation.

Layer 1: X.21 and X.21bis X.21 operates at the OSI Physical Layer. It defines a DTE/DCE interface along the lines of the RS-232 (V.24) standard, except that X.21 was designed for interfacing to a digital network (such as ISDN). Because digital networks were not generally available when X.25 was developed, X.21bis (essentially RS-232) was defined as an interim standard.

PAD For an application to transmit data across an X.25 network, the application's network node must have an X.21 or X.21bis interface, and execute processes that provide the LAPB and X.25 PLP services to the Transport Layer. When X.25 was developed, many devices, such as word processors or "dumb" terminals, did not have these components.

To allow these devices to connect to public X.25 networks, ITU-T developed a set of standards to provide access for terminals and DTE that cannot execute the layers of X.25. The standards, informally called the Interactive Terminal Interface (ITI) standards, are X.3, X.28, and X.29.

The ITI standards collectively define a "black box" called a packet assembler/disassembler (PAD). A PAD accepts a stream of bytes from an asynchronous DTE (such as a PC), "assembles" those bytes into X.25 packets, and transmits the packets on the X.25 network. Of course, the PAD performs the reverse operations for data sent back to the DTE. The Packet Assembler/Disassembler Diagram illustrates the concept of a PAD.

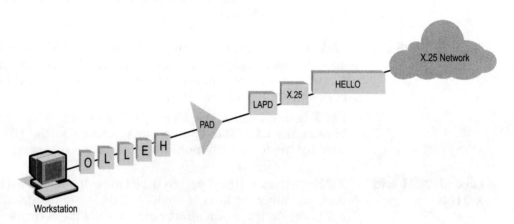

Packet Assembler/Disassembler

To the DTE, the PAD looks like a modem. This means no special software or hardware must be added to the DTE beyond that needed for ordinary asynchronous communications. It is also possible to attach the DTE to the PAD with a point-to-point link using modems. A single PAD can serve several DTE, performing a concentrator function by placing data from more than one DTE into a packet when possible.

PLP

PLP uses two basic types of packets, shown on the Packet Layer Protocol Formats Diagram:

- Data packets
- Control packets

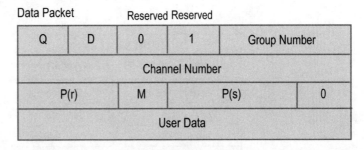

Packet Layer Protocol Formats

The fields for both of these types of packets are listed below:

- **Q bit**—Distinguishes between control and user data information. When the bit is set to 1, the packet contains user data; a bit set to zero indicates a control packet.

- **D bit**—Indicates end-to-end acknowledgment of packets.

- **Reserved**—The 2 bits following D and Q are currently not used.

- **Group number**—Contains the logical channel group number.

- **Channel number**—Identifies the logical channel number. Together, the channel number and group number form the packet address.

- **P(r)**—Contains the sequence number of the next packet to transmit.

- **P(s)**—Contains the value of the packet sent.
- **M bit**—"More data" bit. When set, it means additional related packets are on the way.
- **Packet type**—The command or instruction contained in a control packet. The PLP Control Packet Types Table illustrates the different packet types that can be used in the X.25/LAPB protocols.

Call Setup and Clearing										
DCE to DTE	DTE to DCE	Control Field Value								
Incoming Call	Call Request	0	0	0	0	1	0	1	1	
Call Connected	Call Accepted	0	0	0	0	1	1	1	1	
Clear Indication	Clear Request	0	0	0	1	0	0	1	1	
DCE Clear Confirmation	DTE Clear Confirmation	0	0	0	1	0	0	1	1	
Data and Interrupt										
DCE to DTE	DTE to DCE	Control Field Value								
DCE Data	DTE Data	X	X	X	X	X	X	X	1	
DCE Interrupt	DTE Interrupt	0	0	1	0	0	0	1	1	
Confirmation	Confirmation	0	0	0	1	0	0	1	1	
Flow Control and Reset										
DCE to DTE	DTE to DCE	Control Field Value								
DCE RR (Mod 8)	DTE RR (Mod 8)	X	X	X	0	0	0	0	1	
DCE RR (Mod 128)	DTE RR (Mod 128)	0	0	0	0	0	0	0	1	
DCE RNR (Mod 8)	DTE RR (Mod 8)	X	X	X	0	0	1	0	1	
DCE RR (Mod 128)	DTE RR (Mod 128)	0	0	0	0	0	1	0	1	
Reset Indication	Reset Indication	0	0	0	1	1	0	1	1	
DCE Reset Indication	DTE Restart Confirmation	0	0	0	1	1	1	1	1	
Restart										
DCE to DTE	DTE to DCE	Control Field Value								
Restart Indication	Restart Request	1	1	1	1	1	0	1	1	
DCE Restart Confirmation	DTE Restart Confirmation	1	1	1	1	1	1	1	1	

PLP Control Packet Types

Control packets are used to establish, conduct, and end an X.25 session. The X.25 Packet Sequence Diagram shows a typical packet-exchange sequence required to set up an X.25 connection and transmit data.

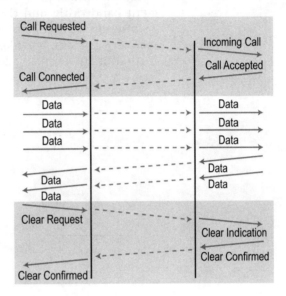

X.25 Packet Sequence

Overhead and Performance Limitations

Faster packet/cell networks, such as frame relay and ATM are replacing X.25 because of the following limitations of the technology:

- **Low throughput**—X.25 networks support DS0 bandwidth at best.

- **High overhead**—Because X.25 PLP is responsible for error-free delivery of packets, each X.25 node in a virtual circuit, plus the Transport Layer process in the receiving node, must acknowledge each packet received. In addition to every "real" data packet that traverses the network, several acknowledgement packets must also make the trip. The result: effective throughput is far lower than the rated capacity of the physical links composing the network.

- **Redundant functions**—X.25's overhead was justified when the public telephone networks were slower and largely analog, as they were when X.25 was first introduced in 1976. However, today's digital networks, which are increasingly based on optical fiber, are much more reliable and have sufficient bandwidth, and thus congestion is not likely to occur. As a result, flow control at Layer 3 is not required, and error recovery can simply be left to the higher layers that must perform it in any event.

Activity

Given the following diagram, describe the movement of information from Client A to Client B and back. In your discussion, describe what protocols are used at each device and across the WAN.

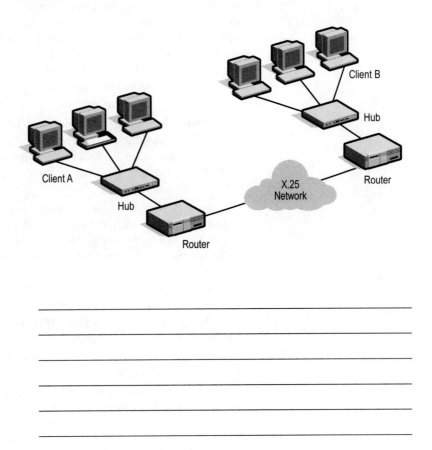

_____

_____

_____

_____

_____

_____

Extended Activity

Given the X.25 Packet Sequence Diagram presented in this lesson, draw X.25 packets as they traverse across a WAN. Assume IP packets are being transported by the X.25 packets. Show which X.25 packets would be carrying IP information and which would be control packets.

Summary

Module 4 began by reviewing the concepts of circuit and packet switching. Circuit-switched protocols, like the public telephone system, are connection-oriented. They use two or more switching devices to set up a virtual circuit across a network. All data in the same transmission uses the same virtual circuit, and the circuit does not change for the duration of the transmission. In contrast, packet and cell networks are connectionless. In those networks, data from a single transmission can take different, changing, often parallel paths across the network "cloud."

SLIP and PPP are widely used by ISPs to provide dial-up Internet connections for home and business users. Both of these simple protocols encapsulate IP packets within frames, and transmit them to an ISP over a local telephone connection. PPP also provides multiprotocol support, data compression, host configuration, and link setup.

ISDN is designed to serve multiple purposes: voice, data, video, facsimile, and all other forms of electronic communication. It uses T-carrier multiplexing technology to provide multiple-channel transmission controlled by out-of-band signaling. ISDN-BRI offers two data-bearing channels (B channels) of 64 Kbps each, plus a control channel (D channel) of 16 Kbps. ISDN-PRI offers 23 B channels of 64 Kbps each, and 1 D channel of 64 Kbps. When necessary, groups of ISDN channels can be bonded, or combined, to form a single large channel.

ATM transports data using small, fixed-length cells, rather than large, variable-sized frames used by T-carrier technology. It provides T3 and SONET speeds, and is gaining widespread acceptance in LAN, MAN, and WAN environments. ATM's unique combination of cell switching and connection-oriented transmission allows it to provide good QoS for both bursty data traffic and time-sensitive multimedia applications.

Frame relay has overtaken X.25 as the packet-switching protocol of choice, because it provides lower overhead and higher speeds (up to T1) than X.25's DS0 data rate. While frame relay is connection-oriented, its variable-length frames make it better suited to data transfers than voice or video.

SMDS is a connectionless, cell-switched service. Like ATM, it uses 53-byte cells that each carry a 48-byte payload; however, SMDS uses a different format for its 5-byte cell header. Its data rate (T1 to T3) takes over where frame relay leaves off. However, SMDS is not a good choice for WANs that cross LATA boundaries, because federal law limits SMDS to metropolitan area service only.

X.25 is important to understand, because it was the first connectionless Network Layer protocol. It uses a telephone number as a destination address, so that switches can route traffic via multiple paths. Newer packet protocols, such as frame relay, were based on X.25 concepts. Although still used to switch packet traffic over a wide area, X.25 is quickly being replaced by protocols that offer faster throughput and lower overhead.

Module 4 Quiz

1. What is the starting justification point to recommend ISDN-PRI?

   a. 6 to 10 lines

   b. 11 to 25 lines

   c. 36 to 45 lines

   d. 26 to 35 lines

2. Which service provides calling party identification in a PBX?

   a. E&M signaling trunks

   b. DDS

   c. Ground start CO trunks

   d. ISDN-PRI

3. A company has geographically dispersed locations and is growing rapidly. The number of locations is growing, as well as the overall traffic, although at different rates for the various sites. Which WAN alternatives could be implemented to address this type of situation? (Choose two.)

   a. Frame relay

   b. ISDN

   c. T1

   d. PPP

   e. SMDS

4. At what point is ISDN-PRI justified?

   a. 1 to 3 lines

   b. 6 to 10 lines

   c. 11 to 25 lines

   d. 36 to 40 lines

5. A business has two locations connected by a point-to-point T1. Although they do not receive busy signals when using T1 for telephone service, they are experiencing slow transfers of data. What can be done to increase the data communications?

   a. Move at least one of the channels from voice to data

   b. Add a DSL line to one of the T1 channels

   c. Increase the bandwidth of the DSU/CSU

   d. Upgrade the T1 router

6. A company has five locations connected by point-to-point T1s. They are experiencing blocked telephone calls between sites. Which actions could be taken to alleviate this situation? (Choose two.)

   a. Upgrade each location to higher-speed T1s

   b. Add another T1

   c. Add PBX ports at each location

   d. Add analog lines to one or more of the locations

7. Which of the following technologies is considered the lowest in the protocol stack?

   a. Packet switching

   b. X.25 packet switching

   c. Frame relay

   d. SONET

8. A cell differs from a packet and frame in which of the following ways?

   a. A cell is much slower than a frame or packet.

   b. A cell is used in connectionless networks.

   c. A cell can be used to transfer analog information.

   d. A cell is a fixed-length entity.

9. A packet is found at which layer of the protocol stack?

   a. Physical

   b. Data Link

   c. Network

   d. Transport

10. A frame is found at which layer of the protocol stack?

    a. Physical

    b. Data Link

    c. Network

    d. Transport

11. Which of the following technologies would not be considered for voice communication between several sites?

    a. TCP/IP

    b. Frame relay

    c. ATM

    d. Leased lines

12. Which of the following protocols is considered Physical Layer only?

    a. ATM

    b. T1

    c. IP

    d. Frame relay

13. Which of the following protocols was initially designed for multimedia?

    a. Frame relay

    b. T1 and T3

    c. ATM

    d. SONET

14. A FRAD might refer to:

    a. Device used for sending voice technology over a frame relay network

    b. Router used to connect a LAN to another LAN using frame relay

    c. Switch that interfaces ATM to frame relay

    d. Frame relay application

15. Frame relay is always more efficient than T1. True or False?

16. ATM, SONET, T1, and frame relay can be used within the same network. True or False?

17. T-carrier installations continue to increase rapidly. True or False?

Module 5
Traffic Engineering

Traffic engineering is an essential part of designing efficient telecommunications systems. In its simplest sense, traffic engineering describes a set of methods for determining the optimum capacity of a communications system, given an accurate description of the traffic load. In other words, once we know the volume and patterns of incoming and outgoing calls, we can use traffic engineering techniques to estimate the number of telephone system components (trunks, switch ports, agents, etc.) needed to support that traffic.

Traffic engineering uses scientific and mathematical methods, such as probability and statistical sampling. However, it is not an exact science. Like weather prediction and opinion polling, traffic engineering can only provide estimates and approximations of the truth. Thus, traffic engineering is a very useful and important first step in the design or analysis of a telephone network. It can help ensure that our first system designs are fairly close to requirements. However, once a telephone network is in daily use, its administrators must analyze actual performance data to refine its design.

Lessons

1. Traffic Engineering Concepts
2. Traffic Data Sources
3. Busy-Hour Engineering

Terms

Bias—Bias is the distortion of scientific sampling results, caused by factors other than those being studied.

Blocked Call—A call that cannot be completed is referred to as a blocked call. Incoming blocked calls receive a busy signal. Outgoing blocked calls must wait for an idle trunk.

Bouncing Busy Hour—The bouncing busy hour is the period of each day when a telephone system carries its greatest volume of traffic. The busy hour "bounces" because the busiest hour usually varies according to the day of the week.

Busy Hour—The busy hour refers to the period of each day, week, or month, during which a telephone system carries its greatest volume of traffic. It is also called constant busy hour.

Call Usage Rate (CUR)—CUR is the time a PBX spends processing calls. See PUR.

Centi Call Seconds (CCS)—CCS also stands for hundred call seconds. One CCS is equivalent to 100 seconds holding time on any number of trunks. Thirty-six CCS is equivalent to one Erlang.

Erlang—One Erlang equals one hour of holding time on any number of trunks. One Erlang is equivalent to 36 CCS. This unit of measurement was named for A.K. Erlang, a Danish telephone engineer who developed a method of estimating the capacity of telephone switches and trunk groups.

Erlang B, C—Erlang B and C are traffic engineering formulas that determine the number of trunks required to serve a particular volume of call traffic at a given grade of service. Erlang B assumes all blocked calls disappear, to redial later or not at all. Erlang C assumes all blocked calls enter a queue, and wait indefinitely until served.

Foreign Exchange (FX)—An FX is a trunk service that lets businesses in one city operate in another city by allowing customers to call a local number. The number is connected, by means of a private line, to a telephone number in a distant city.

Grade of Service—The probability that an incoming call will be blocked during the busy hour is referred to as the grade of service. For example, a grade of service of 0.001 means that, on average, one call in one thousand will be blocked (receive a busy signal).

Holding Time—The holding time is the total time a call uses, or "holds," a trunk.

Molina—The term Molina is a name sometimes used for the Poisson traffic engineering formula and tables. The traffic formula and associated tables were actually developed by E.C. Molina. However, he gave full credit to S.D. Poisson, who had discovered the mathematical principle in the 1820s.

Poisson—Poisson is the traffic engineering formula that determines the number of trunks required to serve a particular volume of call traffic at a given grade of service. Poisson assumes all blocked calls immediately redial, and keep redialing until they get through. The formula is named for S.D. Poisson, who first developed the mathematical approach in the 1820s; it is also called the Molina formula.

Port Usage Rate (PUR)—PUR is the number of PBX ports required to process a call. PUR is typically twice the CUR, because a simple call requires a minimum of two ports (input and output). See CUR.

Private Branch Exchange (PBX)—A PBX is a sophisticated business telephone system that provides all the switching features of a telephone company's CO switch. Today's PBXs are fully digital, not only offering very sophisticated voice services, such as voice messaging, but also integrating voice and data.

Queue—A queue is a collection point where calls are held until an agent or attendant can answer them. Calls are ordered as they arrive and are served in that order. Depending on the time delay in answering a call, announcements, music, or prepared messages may be employed until the call is answered.

Sampling—Sampling is the scientific technique of describing the characteristics of a population by examining only a portion of the population.

Spread—The range between the largest and smallest values in a scientific sample is referred to as the spread. When the range of a sample is close to its average, the sample is more likely to accurately represent the total population being studied.

Tie Line, Tie Trunk—A tie line (tie trunk) is a dedicated circuit that links two points without having to dial a telephone number. Many tie lines provide seamless background connections between business telephone systems.

Trunk—A trunk is a transmission path that connects two telephone switches: two COs, a CO and a business PBX, or two PBXs.

Wide Area Telecommunications Service (WATS)—WATS is a discounted long distance service. Customers can purchase incoming and outgoing WATS separately.

Lesson 1— Traffic Engineering Concepts

This lesson introduces the main traffic theories and principles that help ensure telecommunications networks have enough capacity to serve the needs of their customers. In practice, telecommunications engineers use these theories and techniques to determine the optimum capacity of both switches and trunk groups, for both inbound and outbound traffic. For simplicity, this lesson concentrates on the process of provisioning trunks to handle inbound calls.

Objectives

At the end of this lesson you will be able to:

- Explain why probability is used in traffic engineering

- Name and describe the three dominant traffic theory formulas

- Use traffic measurements to estimate the required number of trunks

- Use traffic measurements to calculate grades of service

 **Key Point**

One goal of traffic engineering is to estimate the optimum number of trunks.

Carried Load

Before we can estimate the required number of incoming trunks, we must know how much traffic to expect, or the carried load. Carried load is the amount of traffic a server or private branch exchange (PBX) actually carries during a given time period. The carried load is limited by the capacity of the PBX and number of trunks available. In other words, either switch capacity or trunk capacity can become a bottleneck that reduces the carried load.

423

Each call uses the full resources of one trunk for a period of time. In a system with multiple trunks, calls start and stop at different times on different trunks. However, no two calls can seize the same trunk at the same time, as shown on the Trunk Utilization Diagram.

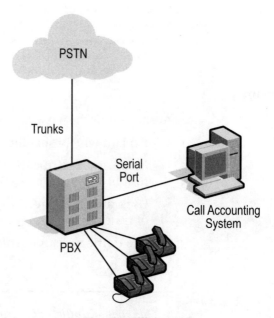

Trunk Utilization

To estimate a carried load, we must know two things:

- Duration, or "holding time," of telephone calls
- Number of incoming calls

There is more to this, of course. However, for now, let us introduce the terms and concepts used to express the number and duration of incoming calls.

Conversation Time

Conversation time is measured from start to finish of a conversation. It starts when a called party picks up the telephone and ends when one party hangs up. AT&T bills standard long distance for conversation time only. This may or may not be the case with other vendors.

Operating Time

Operating time is the time necessary to establish and release a connection with a called party. It includes all necessary dialing and waiting for the called party to answer. There are many factors that influence operating time:

- **Ringing time**—How long it takes a called party to answer a call can affect operating time.

- **Type of equipment**—Different types of PBXs and central office (CO) equipment process calls at different speeds.

- **Human factors**—Manual tie lines require that calls be placed by a PBX attendant.

The numbers and variables possible in operating time make it impossible to establish a standard operating time for all cases. Operating times will always vary between clients and services. Some general rules of thumb are:

- **Dial access (caller dials directly)**—25 to 30 seconds

- **Manual access (PBX attendant places call)**—80 seconds

- **Incoming calls on Wide Area Telecommunications Service (WATS) or foreign exchange (FX)**—20 seconds

Holding Time

In the context of traffic engineering, "holding time" does not mean the time a caller waits on hold. Instead, holding time is the total time one call uses, or "holds," a trunk. The holding time for one call is the sum of its operating time and conversation time; thus, even an unanswered call has a holding time equal to its operating time.

There are three types of holding time of particular concern to telephone system design:

- **Average holding time per call**—The sum of total conversation time and operating time in one time period, divided by the number of calls in the period

- **Hourly holding time**—The total operating and conversation time in a one-hour period

- **Daily holding time**—The total operating and conversation time in one day

Averages are very useful; however, they can sometimes fail to accurately describe call traffic. Therefore, two terms are used to describe the way holding time varies in a system: constant distribution and exponential distribution.

Constant Distribution

When there is little to no variation in call length, we say the holding time has a constant distribution. Common examples of calls with constant distribution are noninteractive services, or very simple interactive services, such as:

- Hotel wake-up calls

- Recorded announcement services, such as time and temperature, dial-a-joke, or theater times and showings

- Credit-check authorization calls

Exponential Distribution

When calls involve two live people, or a wider choice of services, holding time can vary widely. Generally, however, most calls are fairly short. If we were to plot a graph of the number of calls for each holding time, we would see that the number of calls falls off sharply as the holding time increases, as illustrated on the Exponential Distribution Diagram.

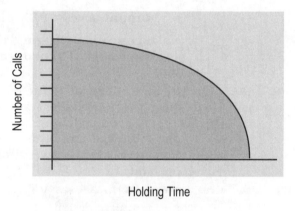

Exponential Distribution

Because this distribution fits the pattern of an exponential formula, it is called an exponential distribution. In other words, the longer the call duration, the fewer number of calls.

Probability

Probability theory is very versatile in communications system design. It is virtually impossible to design a logical system without invoking the theory of probability.

Probability theory is used in many walks of life; to estimate the waiting time for lines at a movie theatre, or determine the number of freeway lanes necessary to carry rush-hour traffic. In telecommunications environments, probability is used to estimate whether a particular level of traffic will create a momentary excess of demand for communication circuits.

The theory of probability expresses the likelihood that an event will occur. As such, probability does not provide exact measurements:

•   The likelihood of the Denver Broncos winning the Super Bowl is 7 to 5.

•   There is a 40 percent chance of rain today.

The probability of an event is the number of favorable outcomes, divided by the total number of outcomes. Probability can be expressed by the following equation:

$$P = f \div (f + u)$$

Where P is probability, f is the number of favorable outcomes, and u is the number of unfavorable outcomes.

We will discuss probability more in the next lesson. For now, let us examine some of the practical ways probability is used to estimate telephone system capacity.

GoS

In a telephone system, grade of service (GoS) expresses the probability that any single call will be blocked. For example, imagine that you call the same business many times. If each of your calls go through on the first try, then the system offers a high grade of service. As a customer, you are very pleased.

If some of your calls get a busy signal, those calls are "blocked" or "lost" calls; they simply cannot get through, because all of the trunks into the system are busy. If those calls go through on subsequent tries, you are probably still satisfied with the grade of service, although it is lower than the first example.

However, if you can never get through, or only get through occasionally, the grade of service is unacceptably low. If a competing business is available, you will start calling them instead.

Thus, grade of service is formally defined as the probability that calls will be blocked, on first try, during the busy hour. To calculate this probability of blockage, we use the probability formula presented above.

In other words, we divide the number of blocked calls by the total number of offered calls (served calls plus blocked calls). For example, imagine that during the busy hour, 5 calls receive a busy signal on the first try, while 254 calls go through on the first try:

Grade of service = 5 ÷ (254 + 5) = 5 ÷ 261 = 0.019

This grade of service means that 19 out of every 1,000 calls, or 1.9 percent of calls, will generally be blocked on first try during the busy hour.

Trunk Usage

We do not normally use minutes and hours in communications engineering. Instead, we use two terms that express time in terms of trunk usage. These two terms measure the carried load of a telephone system.

CCS

The basic measurement of carried load is a centi call second (CCS), short for 100 call seconds (C represents the Roman numeral 100). CCS is sometimes pronounced "cent call seconds." One call second represents one second of holding time on one trunk. Therefore, 1 CCS is equivalent to 100 seconds holding time on one trunk, 1 second holding time on each of 100 trunks, 10 seconds holding time on each of 10 trunks, and so on. For example, a five-minute telephone call is equivalent to 300 seconds holding time on one trunk, or 3 CCS.

If your holding time measurements provide both minutes and seconds, use the following formula to convert to CCS:

(minutes x 60) + seconds/100 = CCS

If you are working with whole minutes, convert to CCS by simply multiplying minutes by 0.6.

For example, a 15.5-minute call is equivalent to 9.3 CCS:

(15 minutes x 60) + 30 seconds = (900 + 30)/100 = 9.3

or

15.5 x 0.6 = 9.3

Erlang

The Erlang is another common measurement of holding time. An Erlang represents one hour of traffic, or 36 CCS:

60 minutes = 3,600 seconds = 36 CCS = 1 Erlang

An Erlang is considered a very large time measurement. It should therefore be carried to the second decimal: 2.54 Erlangs or 0.88 Erlang.

Like CCS, Erlangs measure total carried load. Therefore, 1 Erlang can represent one hour of holding time on one trunk, or 30 minutes of holding time on each of two trunks.

Because an Erlang represents one hour's worth of carried load, it is a convenient way to measure the average number of trunks in use during one hour. For example, if a company has 20 incoming trunks, and measures 15.2 Erlangs of traffic during one hour, then just over 15 trunks were fully busy during that hour, on average.

At first glance, it might seem that four of the company's trunks were wasted. However, using 16 trunks to carry 15.2 Erlangs of traffic would require all of those trunks to be nearly 100 percent utilized. That level of traffic density assumes practically no idle time between calls. To pack that many calls into that number of trunks, each call would have to enter the system just as another call ends.

Of course, that is ridiculous. Telephone calls, like most other human activities, are highly random. Thus, to estimate the number of trunks needed to carry a particular level of traffic, we need to use mathematical tools that model random processes.

The Erlang B Traffic Formula

The Erlang was named after A.K. Erlang, a Danish telephone engineer who was interested in this difficult problem of telephone traffic congestion. In 1918, he published a mathematical technique for estimating the load-carrying capacity of telephone switches and trunk groups.

The Erlang B formula calculates the grade of service likely when a certain volume of traffic is carried by a fixed number of trunks. The formula is based on the following assumptions:

- The number of trunks is limited.

- Calls can originate from an infinite number of sources (from the standpoint of one telephone system, the total number of telephone users is practically infinite).

- Lost, or blocked, calls do not call back; if they do call back, they wait a substantial time before trying.

- The switch can connect any trunk to any inside line. In other words, the switch is "fully available."
- Call holding times are exponentially distributed.

The Erlang B formula itself is beyond the scope of this course. Fortunately, the formula has been used to generate tables that relate the number of trunks to various carried loads and grades of service. With an Erlang B table, we can use any two variables to find the third. We can also perform "what if" analyses with various combinations of values.

The Partial Erlang B Table presents the rows for 10, 11, and 12 trunks. The column for each grade of service measures the size of the traffic load, in Erlangs, that each number of trunks can carry.

Partial Erlang B

Trunks	Grade of Service (Calls Blocked)					
	0.001 (1 per 1,000)	0.002 (1 per 500)	0.005 (1 per 200)	0.01 (1 per 100)	0.02 (1 per 50)	0.05 (1 per 20)
10	3.09	3.43	3.96	4.46	5.08	6.22
11	3.65	4.02	4.61	5.16	5.84	7.08
12	4.23	4.64	5.28	5.88	6.62	7.95

Assume a business has 10 incoming trunks, which carry 4.5 Erlangs of traffic during the busy hour. To find the system's grade of service, first find the 10-trunk row, then read across to locate that level of traffic in Erlangs. In the table, you can see that this system provides a grade of service of 0.01 (1 blocked call per 100 offered calls).

Now let us say that the business wants to increase its grade of service to 0.002, that is, it wants only 1 call in 500 to be blocked. To see how many trunks are necessary to achieve this goal, find the column for the desired grade of service, then read down to find the traffic level. The figure that is closest to the current carried load is 4.62 Erlangs, which is on the row representing 12 trunks (you cannot install a fraction of a trunk, thus always round up). Therefore, to meet its grade of service goal, the business must add two more trunks.

Erlang C and Poisson Tables

By now, you may have noticed a problem with the Erlang B formula. It assumes blocked calls go away, to return later or not at all. Because the real world rarely works that way, two other formulas are also used to estimate trunk capacity under different conditions.

Poisson: Immediate Redial

The Poisson formula, and its associated tables, was actually developed by E.C. Molina, an AT&T engineer working at the same time as Erlang. When Molina discovered that the same mathematical approach had first been discovered around 1820 by S.D. Poisson, he gave the prior researcher full credit. Thus, the Poisson formula is sometimes called the Molina formula (especially within AT&T).

By any name, the formula assumes blocked calls do not go away. Instead, blocked users immediately redial, and keep dialing until they get through. Because this effectively increases the volume of offered calls, a Poisson table tends to estimate a higher number of required trunks than Erlang B.

Erlang C: Infinite Queue

The Erlang C formula assumes all blocked calls enter a queue and wait indefinitely for service. Thus, Erlang C is commonly used to design Automatic Call Distributor (ACD) and call center applications.

The Erlang C formula assumes calls do not leave the queue before being served. Thus, the total offered traffic must be less than the total trunk capacity; otherwise, the queue will grow until it becomes infinitely large.

No Formula is Perfect

As you can see, the Poisson and Erlang C formulas are really no better than Erlang B; they just make different assumptions about an ideal world that does not really exist. To analyze processes as complex as random telephone traffic, mathematicians must simplify the factors; otherwise, the problem would have too many variables to solve.

Therefore, no traffic engineering formula is perfect. To develop a usable design, a telecommunications professional may use different tables to estimate the capacity of different system components.

Estimating Total System Capacity

As stated at the beginning, this lesson simplifies the process of traffic engineering by focusing only on inbound calls. However, outbound calls need trunk capacity too, as well as internal traffic carried over interoffice tie trunks. Furthermore, each key component of the phone system, such as a voice mail system, has its own capacity limitations that must be considered as part of the overall traffic engineering problem.

Outbound Trunk Capacity

Outbound trunks are estimated using the same techniques described above for inbound traffic engineering. However, it is important to analyze outbound and inbound traffic separately when estimating the total number of trunks for a business system. In other words, estimate inbound trunks based on inbound traffic, and outbound trunks based on outbound traffic. Then add the two estimates to determine the total number of trunks between the PBX and the CO.

Tie Trunks

Tie trunks are also estimated using Erlang and Poisson formulas. Accurate data gathering is important for this type of estimate, because a tie trunk between company locations may carry a varying mixture of inbound, outbound, and internal traffic. Let us examine two common scenarios to see how this works.

In one of the simplest cases, a tie trunk links two locations that operate independently, as illustrated on the Independent Locations Linked by Tie Trunk Diagram. Each location has its own inbound and outbound trunks; customers call Locations A and B by using different telephone numbers. Thus, the tie trunk only carries internal traffic between the two offices.

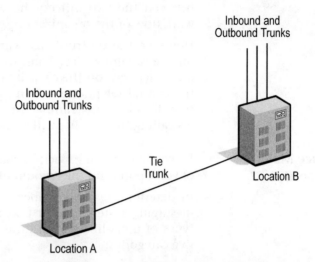

Independent Locations Linked by Tie Trunk

We cannot base a traffic estimate on the number of people in each location, because all workers in one location may not need to communicate with people in the other. Therefore, to estimate the busy-hour traffic flow between Locations A and B, it is best to survey users and analyze the work flow of the organization.

In another common case, all inbound and outbound calls flow through a central location, as illustrated on the Central and Satellite Offices Linked by Tie Trunk Diagram. Tie trunks link the CO to one or more satellite offices, and the CO PBX automatically routes calls to the satellite over the tie trunks.

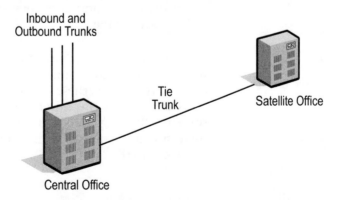

Central and Satellite Offices Linked by Tie Trunk

As in the first example, this tie trunk carries internal traffic between the two offices; the volume of traffic depends on the work flow of the organization.

However, the tie trunk also carries inbound and outbound calls for the satellite office. Thus, the quality of service of the satellite office depends on the capacity of the tie trunks that connect it to the central telephone switching system in the main office location. If too few tie trunks connect the CO to the satellite, customers calling the satellite will frequently receive busy signals.

Voice Mail Capacity

Voice mail system capacity depends on two variables: the number of voice ports and the amount of storage time the system provides.

In determining the number of ports and storage required of a messaging system, the first step is to gather information on the types of users likely to use the system. Voice mail users fall into five categories:

- **Light**—Users who do not heavily use the telephone, and who are likely to answer most or all calls.

- **Medium**—Users who receive many calls, but are still likely to answer them and pick up messages promptly.

- **Heavy**—Users who typically receive more calls than they can immediately answer. These users might leave messages on the system for pickup at a more convenient time.

- **Very Heavy**—These people tend to receive many calls and messages, typically cannot answer some calls as they come in, and do not pick up their messages immediately.

- **Extra Heavy**—People who travel, or whose jobs regularly keep them away from their desks. They cannot answer many of their calls, and keep messages on the system for extended periods of time.

Ports

A voice mail system's ports connect it to the PBX switch. When necessary, the PBX routes an incoming call to the voice mail device over one of these connections. Therefore, the number of ports determines the number of users a voice mail system can support. Blocking, that is, receipt of a busy signal, occurs when there are not enough connections available to carry the call volume into the device.

The Estimated Port Usage Table presents standard guidelines often used to estimate port usage.

Estimated Port Usage

User Type	Port Usage: Minutes Per Day
Light	2
Medium	4
Heavy	6
Very heavy	8
Extra heavy	10

The same Erlang and Poisson formulas used to estimate trunks are also used to calculate voice mail port capacity. The organization decides what grade of service the voice mail device should deliver, then uses the most appropriate formula to determine the number of connections between the voice mail system and PBX.

When estimating traffic, consider heavy traffic created by features such as broadcast and distribution list messaging. For example, when someone sends a 20-second group message to six users, the message creates 120 seconds of traffic.

Storage Time

Voice mail systems must provide storage for both greetings and messages. To prevent the message storage area from filling with unanswered or archived messages, administrators may limit the size of mailboxes.

This storage may consist of either hard disk drives or flash memory. Flash memory can typically only store one to two hours of messages, making it adequate for small two- or four-port systems. Large business systems store hundreds of hours of messages on mirrored hard disks configured for recoverability and reliability. For example, one version of the Avaya Communication Intuity Audix voice messaging system uses 2 gigabytes (GB) of disk space to store 425 hours of speech.

There is no way to exactly calculate the amount of disk space required. However, the following guidelines provide a good starting point, although they will require adjustment over time. The Disk Space Required by Type of User Table lists the estimated amount of disk space, in minutes, for each user type.

Disk Space Required by Type of User

User Type	Basic (Call Answer)	Advanced (Call Answer and Voice Mail)
Light	1.3 minutes	2.0 minutes
Medium	1.9 minutes	2.8 minutes
Heavy	2.3 minutes	3.4 minutes
Very Heavy	2.6 minutes	3.9 minutes
Extra Heavy	3.0 minutes	4.5 minutes

PBX Processing Capacity

As we have seen, both the Erlang and Poisson formulas simplify traffic calculations by making certain assumptions about traffic flow. One of the most important of these assumptions is that the switch is "fully available." In other words, the PBX is a perfect device that immediately connects any incoming call to an inside line or system.

However, if a PBX is overutilized, because of insufficient ports or processing power, it cannot make these instant connections. In other words, it is not fully available. When this happens, the PBX becomes a bottleneck, and the overall system cannot deliver the grade of service predicted by Erlang or Poisson traffic estimates.

Two concepts are used to measure PBX utilization:

- Call Usage Rate (CUR) is roughly equivalent to holding time. CUR measures how long a call has used PBX resources.

- Port Usage Rate (PUR) measures how many PBX ports are used during a call.

PUR more accurately describes PBX utilization, because it essentially measures the usage of time slots across the time-division multiplexing (TDM) backplane of the switch. At a minimum, PUR is twice the CUR, because each call requires at least two ports: one incoming and one outgoing. More complex calls, such as conference calls, may require or use more ports per call.

Therefore, a low CUR does not guarantee a PBX is not the system bottleneck. A system designer should also check the PUR to see how much of a PBX's resources are used to process its current call volume. For example, a low volume of complex conference calls can use up switch capacity just as much as a high volume of simple calls.

Adjusting to Changes

Traffic engineering is an ongoing process, not a one-time job. A telephone system administrator must stay alert to any factors that could require a change in the number or direction of trunks.

A company's traffic patterns change along with its business activity, number of employees, relative size of company locations, and internal work flow. It is obvious that big changes, such as the acquisition of another company, will require a major reestimation of trunk capacity. However, simply transferring workers from one location to another may also have a large effect on interoffice traffic. Therefore, the telephone administrator must work to stay informed of ongoing organizational changes that could require telephone system adjustments.

Administrators can check the effect changes have on system capacity by periodically reviewing traffic reports and reviewing probability calculations. If a new line of business is introduced, either as an addition to an existing system or as an entirely new system, historical data and Erlang or Poisson formulas can provide good predictors of how well the new business systems will perform. The goal is to provide continuing good customer service at a price that allows the business to be profitable.

Activities

1. Traffic formulas are based on _____.

2. The period of each day when there is the most traffic in CCS is:

   a. Erlang

   b. Conversation time

   c. Blocking

   d. Busy hour

3. A call that encounters a busy signal and is therefore not completed is a:

   a. Erlang

   b. Blocked call

   c. Busy hour

   d. Retrial

4. 100 call seconds, the basic measurement of communications engineering, is also referred to as:

   a. CCS

   b. Operating time

   c. Conversation time

   d. Holding time

5. One hour of traffic, a measurement of time used in traffic engineering, is:

   a. Busy hour

   b. Operating time

   c. Erlang

   d. CCS

6. The amount of traffic a server or servers will actually be able to carry during a given busy hour is:

   a. Carried load

   b. Conversation time

   c. Retrial

   d. Erlang

7. The portion of a call, from start to finish, of conversation is the:

    a. Operating time

    b. Holding time

    c. Conversation time

    d. Carried load

8. The percentage of blocked calls that retry the server is the:

    a. Carried load

    b. CCS

    c. Erlang

    d. Retrial

9. The time necessary to establish and release a connection with the called party is the:

    a. Operating time

    b. Conversation time

    c. Carried load

    d. Busy hour

10. The sum of operating time and conversation time is the:

    a. Operating time

    b. Holding time

    c. Busy hour

    d. Blocking

11. The inability to complete a call due to lack of facilities is referred to as:

    a. Busy hour

    b. Holding time

    c. Blocking

    d. Erlang

Extended Activities

1. What is the grade of service given the following information for a company's busy hour?

   8 calls receive a busy signal

   376 calls go through on the first try

2. What is equivalent to the following in CCS?

   a. 18-minute call

   b. 3-minute call

   c. 1.5-minute call

3. What is the grade of service for a business telephone system that has 11 trunks and 4 Erlangs of traffic?

Lesson 2—Traffic Data Sources

As we saw in Lesson 1, the major traffic engineering formulas require us to know the volume of incoming traffic during the busy hour. Therefore, to estimate trunk capacity and grade of service, we need accurate measurements of actual traffic.

Just as there is no single perfect traffic estimation formula, there is no single right way to gather telephone traffic data. This lesson discusses some of the factors to be aware of as we collect measurements necessary for our design.

Objectives

At the end of this lesson you will be able to:

- Explain the purpose of sampling
- Describe unintentional bias
- Name and describe the two keys to sampling accuracy

 **Key Point**

To provide accurate answers, traffic data must be carefully gathered.

Data Sources

There are many rich sources of traffic data, including:

- Carrier call detail reports
- Long distance billing
- PBX attendant tickets
- Traffic-related data from smart multiplexers (MUXs)
- Station Message Detail Recordings
- Diagnostic systems
- Traffic monitors on tie lines
- Traffic information from ACDs
- WATS detail records
- Customer records of business volume, order volume, etc.

The type of data you need will determine the data sources you choose. For example, to calculate grade of service, you must know both the number of calls that get through the PBX, and the number of calls blocked because all trunks are busy. Because a PBX never "sees" blocked calls, blocked-call statistics are only available from the switch that attempted to send them, usually the telephone company CO.

Gathering Data

Each PBX design situation requires different types and quantities of data. There are, however, three key factors to consider in every data-gathering situation:

- Accuracy of collected data
- Cost of collecting the data
- Time needed to collect the data

Data accuracy considerations can be further subdivided:

- How accurate will the data be?
- How accurate should the data be?
- How sensitive is the design to inaccuracies in the basic data?

Both the accuracy and cost of the data collected in any particular case will involve some, or all, of the following factors:

- Type of data to be collected
- Amount of data to be collected
- Personnel and equipment required, and whether they are already available
- Method of data collection

When considering the accuracy and cost of collecting your data, bear in mind how you will analyze the data. You must consider how expensive data collection will be and how much time will be required to analyze the collected data. The time necessary to collect data may also be a critical issue, and is influenced by the four factors listed above.

It is evident that the accuracy, cost, and time involved in collecting traffic data are highly interrelated. The basic question is, therefore: How much data should be collected? To answer that question, we must return to probability theory.

Experiments

An experiment is a controlled process of observation. For example, an experiment can consist of flipping a coin 100 times to determine the probability of getting heads vs. tails. Each experiment records the occurrence of an event. Thus, when flipping a coin, we measure how many times the coin comes up heads (or tails), and compare that to the total number of flips.

Some events are mutually exclusive; that is, they result in a yes or no answer. A coin can either be heads or tails, not both. Other events are variable. For example, the duration of telephone calls can be short, long, or somewhere in between.

The theory of probability is always used to express the results of an experiment. If we flip a coin 100 times, we will probably discover that it comes up heads about 50 times. However, that knowledge does not tell us whether any individual coin toss will come up heads. All the experiment tells us is that there is a 50 percent probability that any individual coin toss will come up heads.

When working with experiments, there are two important factors to consider:

- Experimental conditions
- Number of experiments

Experimental Conditions

The outcome of an experiment can be heavily influenced by the conditions under which it is performed. For example, a few of the experimental conditions that can alter the results of the coin-flipping experiment are as follows:

- **Coin**—Is it balanced?
- **Personnel**—Does the person flip the coin so that one side or the other comes up more often?
- **Environment**—Does the air flow, temperature, or some other condition affect the outcome?

While some of these variables may seem far-fetched, experimental results have been thrown off by factors just as small. Therefore, scientists must often go to extreme lengths to eliminate all influences except the ones they are trying to study.

Number of Experiments

The more events an experiment measures, the less important each event becomes to the overall outcome. Thus, when determining a probability value, it is important to perform a large enough number of experiments.

The percentage of heads in 1,000 coin flips is more representative than 10 flips. The larger the number of events, the more likely the data will represent a realistic picture. That is because a large number of experiments tends to cancel out the effect of any coincidences, or wildly variable data.

Sampling

As an event becomes more complex, it becomes harder to create an impartial experiment that models it. This is especially true of any event that includes human interaction. Thus, it is often easier and more accurate to measure real-world events.

There are two basic techniques of measurement:

- Examine all items

- Examine some of the items

Examining all items is the most accurate type of measurement, but it is expensive and time consuming. For example, the U.S. Census collects information about every individual citizen and detailed information about every sixth household; however, this intensive study is so difficult it is done only once every 10 years.

Fortunately, we usually do not need to examine every event to get a clear understanding of the real world. Sampling is the scientific technique of describing or estimating the characteristics of the population from an examination of only a portion of the data collected. For example, if we want to know what portion of the population is left-handed, we can examine a portion, or random sample, of people, then apply that percentage to the overall population.

Bingo is an example of random sampling. Balls are tumbled in a large cage, so that each has the same probability of being chosen. It is fairly easy to ensure the randomness of a simple system such as bingo balls; however, real-world random sampling is much more difficult, because many factors may reduce the randomness of a sample.

Bias

Just as experimental results can be skewed by uncontrolled influences, the results of a random sample can be altered by the characteristics of the items being studied. This effect is called unintentional bias, and it can significantly affect the results of a study. For example, if we are trying to determine the proportion of left-handed people in the population, it makes no sense to choose our random sample from attendees of a left-hander's convention.

A classic example of unintentional bias occurred in the 1936 U.S. presidential election. A Literary Digest poll mailed more than 10 million ballots to households listed in telephone books and automobile registration records. The poll's results predicted Alf Landon as the winner over Franklin Roosevelt. What the poll did not take into consideration was that millions of Depression-era Americans could not afford either telephones or cars, and those millions voted for Roosevelt.

There are two keys to producing a reliable random sample:

- Sample size
- Sample spread

Sample Size

The numerical size of a sample, not its percentage of the whole, is the most significant factor in its reliability. For example, a database of 100,000 calls in which 1,000 are studied equals a 1% sample. A sample of 1,000 studies out of 10,000 calls is equivalent to a 10% sample. A sample of 200 studies out of 1,000 calls is equivalent to a 20% sample. The 1% and 10% studies will be the most accurate (approximately 96%), because they are large enough to reduce the effects of individual elements. A sample of only 200 calls is too small, and probably will not be accurate enough.

In traffic measurement, a one-hour sample is too brief and probably does not provide a true picture of reality. A five-day study is more likely to provide reliable data, because it includes enough data for skews to settle down. To determine one day's usage, average the data gathered over five days. In general, gather as large a sample as your budget can justify.

Sample Spread

After you have the proper sample size, it is important to determine the spread, or range between the largest and smallest items in the sample. For example, assume your analysis of a sample determines that the average call length is five minutes, and 95 percent of all calls range between four and six minutes. With this type of information, you have a very high probability that your sample is accurate because there is a tight cluster around the average.

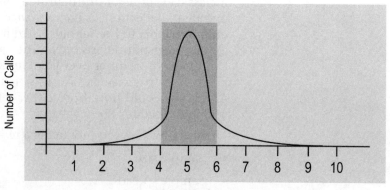

Holding Time (minutes)

On the other hand, if only 40 percent of your calls fall between four and six minutes, there is a very high probability that a five-minute average call length is not representative of actual traffic. If this is the case, very long and very short calls should also be considered in the study. The data must be analyzed very carefully and/or a larger sample should be taken.

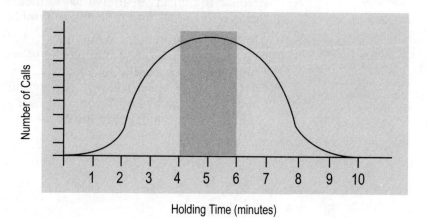

Holding Time (minutes)

Planning a Study

Before you begin collecting data, or choose past data to analyze, it is important to clearly define the goal of your design. Your design goal is essentially a question that your traffic study must answer.

Thus, if your goal is a system with enough capacity for typical day-to-day business activities, your traffic study must define exactly what happens on a typical business day. This means you must eliminate as many factors as possible that are not typical. Pay attention to your company's monthly billing cycle, annual sales patterns, or typical vacation schedules. Any of these factors could generate telephone traffic that is abnormally high or low.

On the other hand, your goal may be a system with enough capacity to provide high service quality during even the heaviest periods. In that case, plan your study to record the heaviest periods of traffic.

Activities

1. Match the following concepts with a description:
   a. Gets higher with the amount of data collected
   b. Controlled process of observation
   c. Example is one flip of a coin
   d. Estimating the characteristics of a population
   e. Skewed results of an experiment
   f. Range between the largest and smallest items in a sample
   g. Number of blocked calls

   Accuracy _____

   Experiment _____

   Event _____

   Sampling _____

   Bias _____

   Spread _____

   Grade of Service _____

2. The purpose of sampling is to _____.

3. Two techniques of sampling are:
   a. Odd and even
   b. Random and scientific
   c. Voice and data
   d. Scientific and unscientific

4. The two keys to sampling accuracy are:
   a. Random and scientific
   b. Expensive and time consuming
   c. Size and spread
   d. Uncertainty and assurance

Extended Activities

1. What is the sample percentage of the following numbers?

    a. 100,000 calls, 20,000 sampled

    b. 500,500 calls, 1,000 sampled

    c. 150,000 calls, 150 sampled

2. For each of the three sample percentages above, answer the following question: If the average length of the calls measured is 4 minutes and 97 percent of the sampled calls fall between 3.8 and 4.2 minutes, would you consider the sample accurate? Explain.

    _____

    _____

    _____

    _____

Lesson 3—Busy-Hour Engineering

The goal of traffic engineering is to provide good service at a reasonable cost. Specifically, we must provide services that can carry busy-hour loads without excessive blocking, yet not provide so much extra trunk capacity that the system's cost is too high.

Objectives

At the end of this lesson you will be able to:

- Explain the difference between a constant busy hour and bouncing busy hour

- Explain how a traffic engineer balances acceptable service and reasonable cost

- List some telephone system components, other than trunks, that can become bottlenecks

 **Key Point**

Traffic engineering must find a balance between customer service and economy.

The Busy Hour

The busy hour is the period of each day, week, or month when a telephone system carries its greatest volume of traffic. To support the activities of a business, a telephone system must have enough capacity to handle a surge of activity. However, to control costs, the system must have no more capacity than is absolutely necessary.

It is important to consider the normal pattern of business activity when analyzing busy-hour traffic. Many organizations have a fairly constant level of activity throughout the year, while others are highly seasonal. For example, a call center for catalog sales will experience heavy traffic from mid-November until just before Christmas. A telephone system designed for mid-summer traffic levels will be totally inadequate to handle the load.

After you have chosen the right time period to study, there are two approaches for determining busy hour:

- Constant busy hour
- Bouncing busy hour

Constant Busy Hour

Constant busy hour is determined by looking at the total traffic volume for each hour of the study period. In the Traffic Volume Table, the traffic for each hour is measured in CCS. If the study period spans several weeks, the traffic volumes for each time slot are averaged separately. For example, all traffic for 9 to 10 A.M. Monday is averaged separately from traffic for 9 to 10 A.M. Tuesday.

Traffic Volume

Time	Traffic Volume (CCS)					
	Monday	**Tuesday**	**Wednesday**	**Thursday**	**Friday**	**Total**
9-10 A.M.	10	12	6	10	14	52
10-11 A.M.	8	10	8	10	6	42
11-12 P.M.	10	12	10	8	10	50
12-1 P.M.	8	12	6	10	8	44

In the table, the busiest hour of the week occurs from 9 to 10 A.M. Friday.

Bouncing Busy Hour

The bouncing busy hour is determined by choosing the busiest hour of each day. Using this technique, the busy hour is very likely to change from day to day. In the above table, 9 to 10 A.M. is the busy hour on Friday, while 11 to 12 P.M. is the busy hour on Wednesday. As you can see, it is also possible to have several busy hours in one day.

Grade of Service vs. Cost

As stated previously, it is important to engineer a telephone system to handle busy-hour traffic. However, it is exorbitantly expensive to create a system that handles 100 percent of a traffic volume that only happens once or twice a year.

To find a balance between customer service and cost containment, we usually engineer for the busiest hour of an average day in the busiest month. This engineering technique means that the telephone network will be overloaded during the busiest hour of the busiest day of the busiest month. However, this practice is generally acceptable, because service does not have to be perfect to be good. The business chooses the grade of service (probability of blocking) acceptable to its customers, then provisions enough trunk capacity to provide that level of service.

This technique also means a business must pay for some unused capacity during the slower times. This is also acceptable, because it is essential that a business provide a decent grade of service during the busy hour.

In this way, telephone trunks are like restaurant tables. A successful restaurant may be packed during lunch and dinner, but it is mostly empty for the rest of the day. However, it would be ridiculous to suggest that a restaurant provide fewer tables, based on an all-day average number of customers. It would also be foolish for a restaurant to provide so many tables and servers that customers never have to wait. Instead, restaurants provide as many tables as they can afford, and customers expect to wait a reasonable time during busy hours.

Potential PBX Bottlenecks

Thus far, we have focused on the traffic engineering goal of providing enough trunks. However, other elements of a telephone system can also become bottlenecks.

Several PBX components are sensitive to the amount of traffic being processed by the system. Faulty traffic engineering can lead to a configuration that contains too many or too few of these components, and subsequently either overprices the switch or degrades its performance. These traffic-sensitive components are:

- Tone detectors
- Processor occupancy
- Time slots
- Modules

Most PBXs can provide statistics that show overall utilization, as well as the load on the central processing unit (CPU) and individual ports. These reports can provide valuable clues when searching for bottlenecks at the PBX.

Tone Detectors   If no tone ports are available to provide dial tone, a call will be blocked. Incoming calls utilizing tone receivers, and data calls utilizing data ports, can be delayed.

Processor Occupancy   The processor time required to handle a given traffic load is called Dynamic Processor Occupancy. The time required to handle various types of traffic depends on a combination of the following factors:

- Application type

- Facility type

- Voice/data terminal type

- Associated software features

Time Slots   Digital telephone switches use TDM to move call signals between trunks and inside lines. Thus, the capacity of a switch is directly related to the number of TDM time slots available. The number of slots, in turn, depends on the number of switching cabinets in the system and the system architecture.

Modules   The final number of modules required in any system design can be driven by hardware port quantities or traffic requirements. Preliminary module estimates in any design are first done based on the number of ports required. Traffic requirements are then calculated to determine additional module requirements.

Consider All Traffic, All Paths

Clients generally specify the number of trunks they want in a configuration. However, alternate routing, such as one-way in and out trunks, FX, WATS, and 800 service, are often overlooked. This often results in overtrunking, and a larger system than is actually necessary.

Therefore, telecommunications engineers must not consider traffic as a total lump sum. Instead, we must consider all paths by which calls can enter a system, and measure the traffic patterns on each of those paths.

Activities

1. Create a table like the Traffic Volume Table "Constant Busy Hour" section of this lesson. Use the same time spans and following numbers in the table:

   9, 9, 7, 14, 6

   8, 8, 7, 14, 7

   9, 9, 7, 15, 7

   8, 9, 8, 14, 7

2. Based on the numbers in the table above, does this organization have a constant busy hour or bouncing busy hour?

   _____

   _____

3. As in Activity 1, use the same time spans and following numbers to create a second table:

   8, 8, 12, 6, 5

   12, 8, 7, 8, 6

   6, 14, 8, 5, 5

   7, 2, 7, 5, 13

4. Based on the numbers in the table above, does this organization have a constant busy hour or bouncing busy hour?

Extended Activities

1. How does grade of service impact a business' cost of telecommunications services?

   _____

   _____

2. What are the tradeoffs between grade of service and cost?

   _____

   _____

3. How can processor occupancy be impacted by applications used by a PBX? How can this affect the ability of a PBX to process incoming and outgoing calls?

   _____

   _____

4. Why must traffic engineers consider all types of trunks when engineering a communications system?

   _____

   _____

Summary

This module briefly introduced the very large and complex subject of traffic engineering. Even under the best conditions, traffic engineering can only provide an estimate of the optimum capacity of a telephone system. However, these estimates are considerably better than guessing. Therefore, traffic engineering should be part of any system design or performance analysis.

To improve the quality of a traffic estimate, a telecommunications professional should pay attention to three issues:

- Choose a traffic formula that models the customer's conditions.

- Gather data that is as accurate as possible.

- Consider all potential paths and bottlenecks, in addition to trunks.

Lesson 1 introduced the three most commonly used traffic engineering formulas, and the tables generated from them. The most-used table, Erlang B, assumes all blocked calls disappear; it tends to deliver a low estimate of the number of trunks necessary to support call traffic. Another popular table, based on the Poisson formula, assumes all blocked calls keep redialing until they get through; this approach tends to estimate a higher number of trunks. The Erlang C formula assumes no incoming calls are blocked; instead, they enter a queue and wait indefinitely for service.

There are other traffic engineering formulas besides these three, and one of those may more closely model conditions at a customer's site. However, all traffic formulas must make assumptions about an idealized set of conditions. Therefore, to get the best estimate, it may be necessary to do "what-if" analysis with more than one formula.

In Lesson 2, we learned that the accuracy of a traffic estimate also depends on the quality of the data fed into each table. CO telephone switches and PBX/ACD systems can provide many different kinds of call traffic statistics. However, traffic engineers must carefully choose the type and amount of data to gather from these systems. In general, you should gather data that closely represents the conditions your system must support, and gather enough data to cancel out the effects of any unusual events.

The concept of busy hour is an important guide when gathering traffic data. As we saw in Lesson 3, a business may base its traffic engineering on one busy hour per month, one per week, several per day, or any other pattern that describes the conditions of the site.

However, when collecting data and analyzing traffic, it is important to consider all paths that can carry calls or provide service. Thus, if some of an organization's calls come in over 800 lines, or from PBXs at branch locations, the traffic flow on those lines should be analyzed separately from other flows. This increases the complexity of the analysis; however, when done correctly, will also increase the accuracy of the resulting estimate.

Module 5 Quiz

1. A telephone call that cannot be completed is referred to as a:

    a. Busy signal

    b. Blocked call

    c. Busy hour

    d. Holding time

2. One hour of holding time on any number of trunks is referred to as:

    a. CCS

    b. Poisson

    c. Erlang

    d. Molina

3. The two traffic engineering formulas used to determine the number of trunks required for handling specific traffic volumes are:

    a. CCS

    b. Poisson

    c. Erlang

    d. Molina

4. The amount of traffic a PBX carries during a given time period is referred to as:

    a. Busy hour

    b. Sampling

    c. Queue

    d. Carried load

5. The time necessary to establish and release a connection with a called party is referred to as:

    a. Conversation time

    b. Operating time

    c. Busy hour

    d. Carried load

6. The traffic engineering principle of "holding time" is the time a caller:

   a. Uses a trunk

   b. Waits on the line

   c. Waits for an agent

   d. Uses an automated attendant

7. Which of the following is not a key factor to consider in data gathering?

   a. Accuracy of collected data

   b. Cost of collecting data

   c. Erlangs

   d. Time needed to collect the data

8. Which of the following would be the most accurate type of measurement?

   a. 10 percent of items

   b. 50 percent of items

   c. 75 percent of items

   d. 100 percent of items

9. If the busiest hour of each day's telephone calls varies throughout the week, this is referred to as:

   a. Constant busy hour

   b. Variable busy hour

   c. Bouncing busy hour

   d. Changing busy hour

10. Which of the following PBX components is not sensitive to traffic?

   a. Tone detectors

   b. Processor occupancy

   c. Time slots

   d. Attendant console

COURSE QUIZ

1. What is the major difference between a LAN and WAN?
   a. Type of cabling used to configure the network
   b. Distance between nodes in the network
   c. Type of devices used to connect workstations to the network
   d. Type of NIC used

2. Which is the fastest line speed?
   a. T1
   b. SONET
   c. 256 Kbps
   d. 512 Kbps

3. Which technology uses a CIR to state the maximum average circuit rate?
   a. X.25
   b. Frame relay
   c. ATM
   d. SONET

4. A PBX becomes necessary when a company grows to:
   a. 5 to 10 lines
   b. 8 to 12 lines
   c. 12 to 18 lines
   d. 20 to 50 lines

5. SLC circuits are used to:

    a. Reduce the number of physical wires

    b. Increase the availability of tie lines

    c. Decrease the chances of power failure and loss of telephone service

    d. Reduce the size of equipment cabinets

6. Which of the following is not a component found in a PBX system?

    a. Trunks

    b. Attendant console

    c. Administrative terminal

    d. DID

7. Incoming calls are connected directly to employees of an organization by means of:

    a. FX

    b. DISA

    c. DID

    d. Attendant console

8. After answering a call, an employee can transfer the call to another extension using which of the following features?

    a. Call forwarding

    b. Call pickup

    c. Speed dialing

    d. Call transfer

9. The PBX feature that allows callers to use the telephone touchpad to access information is known as:

    a. IVR

    b. DID

    c. Administrative dialing

    d. Call processing

10. The system or PBX feature that routes calls to telephone support personnel is referred to as a:

    a. UPS

    b. DSS

    c. ACD

    d. SMDR

11. The feature of a PBX system that creates a log of all incoming and outgoing calls is called:

    a. UPS

    b. DSS

    c. Call accounting

    d. SMDR

12. The PBX feature that allows a user to transfer incoming telephone calls to another telephone extension on the PBX system is:

    a. Call pickup

    b. Call forwarding

    c. Call waiting

    d. DAU

13. The PBX feature that allows a user to answer another telephone by means of his or her telephone is:

    a. Call pickup

    b. Call park

    c. Call waiting

    d. DAU

14. The PBX feature that allows a user to resume a telephone call from another extension is:

    a. Call pickup

    b. Call park

    c. Call waiting

    d. DAU

15. One hour of holding time on any number of trunks is referred to as:

   a. CCS

   b. Poisson

   c. Erlang

   d. Molina

16. The traffic engineering formula used to determine the number of trunks required for handling specific traffic volumes is:

   a. CCS

   b. Poisson

   c. Erlang

   d. Molina

17. The amount of traffic a PBX carries during a given time period is referred to as:

   a. Busy hour

   b. Sampling

   c. Queue

   d. Carried load

18. The time necessary to establish and release a connection with a called party is referred to as:

   a. Conversation time

   b. Operating time

   c. Busy hour

   d. Carried load

19. The traffic engineering principle of "holding time" is the time a caller:

   a. Uses a trunk

   b. Waits on the line

   c. Waits for an agent

   d. Uses an automated attendant

20. Which of the following is not a key factor to consider in data gathering?

    a. Accuracy of collected data

    b. Cost of collecting data

    c. Erlangs

    d. Time needed to collect the data

21. If the busiest hour of each day's telephone calls varies through-out the week, this is referred to as:

    a. Constant busy hour

    b. Variable busy hour

    c. Bouncing busy hour

    d. Changing busy hour

22. Which of the following PBX components is not sensitive to traffic?

    a. Tone detectors

    b. Processor occupancy

    c. Time slots

    d. Attendant console

23. Which of the following describes a simple point-to-point network?

    a. Network that consists of a single type of cable

    b. Network that consists of one type of computer such as a PC or Macintosh

    c. Network where only two computers are communicating at a given point in time

    d. Network where a cable connects many computers

24. A connectionless network is a network where:

    a. No connection is established between endpoints.

    b. No physical path exists between any two points in the network.

    c. A circuit-switched network is used.

    d. All nodes attached to the physical media receive the transmission.

25. The copper cabling that connects many homes and businesses to the first CO is referred to as the:

    a. Local loop

    b. Trunk lines

    c. Digital loop

    d. Leased lines

26. Which of the following types of telecommunications circuits are typically fixed established over a fixed path?

    a. Switched line

    b. Dial-up line

    c. Full-duplex line

    d. Leased line

27. ADSL is best characterized by:

    a. Analog-to-digital conversion at the local loop

    b. Digital-to-analog conversion at the local loop

    c. High speed to the subscriber, low speed from the subscriber

    d. High data transfer from the subscriber, low speed transfer to the subscriber

28. An example of a point-to-point alternative used to transmit data over a wide area might be:

    a. Leased lines

    b. X.25

    c. ISDN

    d. ATM

29. ADSL gets its name from the fact that:
    a. It is an analog service.
    b. It is a permanent service.
    c. Information travels from subscriber to CO only.
    d. Information travels in one direction faster than in another direction.

30. FT1 is a multiple of:
    a. 64-Kbps channels
    b. 58-Kbps channels
    c. T1 channels
    d. T3 channels

31. T1 is equivalent to:
    a. DS0
    b. ISDN Basic Rate
    c. DS1
    d. E1

32. A MUX is used to:
    a. Map low-speed input signals to a high-speed output signal
    b. Map high-speed input signals to a low-speed output signal
    c. Convert analog signals to digital signals
    d. Convert digital signals to analog signals
    e. None of the above

33. The building block for SONET is:
    a. STS-1
    b. 51.84 Mbps
    c. 48 Kbps
    d. 64 Kbps
    e. Both a and b

34. When ATM is carried at 155-Mbps rates, the physical transport is:

    a. DDS

    b. T1

    c. T3

    d. SONET

35. The primary difference between an STS signal and OC signal is:

    a. One is digital and the other is analog.

    b. One is low-speed and the other is high-speed.

    c. One is electrical and the other is optical.

    d. One is binary and the other is octal.

36. A cell differs from a packet and frame in which of the following ways?

    a. A cell is much slower than a frame or packet.

    b. A cell is used in connectionless networks.

    c. A cell can be used to transfer analog information.

    d. A cell is a fixed-length entity.

37. A packet is found at which layer of a protocol stack?

    a. Physical

    b. Data Link

    c. Network

    d. Transport

38. A frame is found at which layer of a protocol stack?

    a. Physical

    b. Data Link

    c. Network

    d. Transport

39. SONET is found at which of the following layers?

    a. Physical

    b. Data Link

    c. Network

    d. Transport

40. Which of the following techniques is the least efficient?

    a. Frame switching

    b. Cell switching

    c. Packet switching

    d. Frame relay

41. All layers of the SONET protocol relate to the Physical Layer of the OSI model. True or False?

42. An ATM cell could consist of multiple SONET frames. True or False?

43. A MUX can take multiple simultaneous digital input streams and put them onto a single digital output stream. True or False?

44. A MUX can take a high-speed digital bit stream and divide it into multiple digital output streams. True or False?

45. A codec performs the same basic operation as a modem. True or False?

46. The physical medium used to transmit information at the highest rates is fiber optic cable. True or False?

GLOSSARY

802.1x—IEEE 802.1x is a recently approved IEEE standard for port-based access control. It is used to control access to a network access device (switch, access point, etc.)

802.3af—IEEE 802.3af is a standard proposed by the Institute of Electrical and Electronic Engineers (IEEE) for powering Ethernet devices over twisted pair cabling. IEEE 802.3af is a legacy Ethernet-compatible, internationally standard power distribution technique.

Adjunct Processor—An adjunct processor is an external computer that controls a PBX. For example, a CTI server is an adjunct processor. In the Avaya DEFINITY PBX, an API called ASAI makes it simpler to write CTI link applications that allow a CTI server to control a DEFINITY PBX.

Adjunct-Switch Application Interface (ASAI)—ASAI is a Lucent-specific API for its DEFINITY PBX switch. ASAI provides commands and messages that enable features such as event notification and call control. CTI link applications that use ASAI can more easily access these services on a DEFINITY system.

Advanced Intelligent Network (AIN)—AIN is the PSTN architecture that relies on centralized network servers to make call routing decisions, rather than leaving these decisions up to the individual telephone switches. SS7 control points run protocols that in turn support AIN features, such as caller ID and call blocking.

Agent—A person or automated device that serves incoming callers is referred to as an agent.

Agent State—Agent state refers to the current availability status of an agent. The term also represents a user's ability to change an agent's availability within the system.

Americans with Disabilities Act (ADA)—The ADA is a federal law that guarantees equal opportunity for individuals with disabilities in public and private sector services and employment. Generally, the ADA bans discrimination on the basis of disability, and requires employers to make "reasonable accommodation" to allow the employment of people with disabilities who are otherwise qualified to perform the essential functions of a job.

Announcement—An announcement is a prerecorded message delivered to a caller in a queue requesting the caller to remain online, prompting the caller for information, or directing the caller to another destination. Announcements can be scheduled to occur in a particular order, or repeat periodically.

Asynchronous Transfer Mode (ATM)—ATM is a connection-oriented cell relay technology based on small (53-byte) cells. An ATM network consists of ATM switches that form multiple virtual circuits to carry groups of cells from source to destination. ATM can provide high-speed transport services for audio, data, and video.

Attenuation—The weakening of a signal over distance is referred to as attenuation.

Auto attendant—An auto attendant is a device, or PBX function, that answers incoming calls, plays a greeting and instructions, then allows callers to directly dial an extension number, or the attendant. This feature can be used to relieve receptionists during peak calling periods and breaks; however, many companies use automated attendants instead of receptionists.

Automatic Call Distributor (ACD)—An ACD is a programmable system that controls how inbound calls are received, held, delayed, treated, and distributed to call center agents.

Automatic Number Identification (ANI)—ANI is a trunk-based feature that passes the number of a calling party to the telephone receiving the call. This information can be sent in several formats, both analog and digital. Some circuits send the information as a series of tones between the first and second ring. Others send the information as a series of DTMF tones after the call is answered. Still other digital circuits (such as ISDN) send this information on a separate digital channel using SS7. Although similar to Caller ID, ANI cannot be blocked by the calling party.

Avaya Access Security Gateway (ASG)—ASG is a suite of products designed by Avaya to secure access to IP and communication systems.

Avaya Communications Manager (ACM)—ACM is the next generation of Avaya call processing software. ACM runs on legacy DEFINITY servers, as well as on Avaya's new line of media servers. ACM is a component of Avaya's Enterprise Class Internet Protocol Solutions (ECLIPS) product line.

Bandwidth—Bandwidth is the total information-carrying capacity of a network or transmission channel. It is the difference between the highest and lowest frequencies that can be transmitted across a transmission line or through a network. Bandwidth is measured in Hz for analog networks and bps for digital networks. See hertz.

Best Service Routing (BSR)—BSR is an ACD feature that allows the ACD to determine which split or skills can provide the best possible service to the caller. BSR can be used in a single site or across multiple sites, where it works in conjunction with LAI to send the call to the best qualified resource (either local or remote).

Bias—Bias is the distortion of scientific sampling results, caused by factors other than those being studied.

Blocked Call—A call that cannot be completed is referred to as a blocked call. Incoming blocked calls receive a busy signal. Outgoing blocked calls must wait for an idle trunk.

Bouncing Busy Hour—The bouncing busy hour is the period of each day when a telephone system carries its greatest volume of traffic. The busy hour "bounces" because the busiest hour usually varies according to the day of the week.

Bridge—A bridge is a device that operates at the Data Link Layer of the OSI model. A bridge can connect several LANs or LAN segments. It can connect LANs of the same media access type such as two Token Ring segments, or different LANs such as Ethernet and Token Ring.

British Thermal Unit (BTU)—A BTU a standard measure of thermal energy. The heat output of electronic equipment is specified in BTUs per hour.

Burst Range—Burst range refers to network traffic outside the range of the CIR.

Bursty—A network traffic pattern in which a lot of data is transmitted in short bursts at random intervals is referred to as bursty.

Bus—A bus connects the central processor of a PC with the video controller, disk controller, hard drives, and memory. Internal buses are buses such as AT, ISA, EISA, and MCA that are internal to a PC but not "local buses."

Busy Hour—The busy hour refers to the period of each day, week, or month, during which a telephone system carries its greatest volume of traffic. It is also called constant busy hour.

BX.25 Protocol—BX.25 is a Lucent/Avaya proprietary variation on the X.25 packet-switching protocol. BX.25 is used in Lucent Distributed Communications System and Digital Communications Interface Unit links.

Cable Vault—A cable vault is a below ground box, roughly 4 x 8 feet, that provides maintenance access to buried cable. It is also called a "maintenance hole."

Call Center—A call center provides a centralized location where a group of agents or company representatives communicate with customers by means of incoming or outgoing calls.

Call Coverage—Call Coverage provides automatic call redirection, based on specified criteria, to alternate answering positions in a Call Coverage path. A coverage path can include a telephone, an attendant group, an Automatic Call Distribution (ACD) hunt group, a voice messaging system, or a Coverage Answer Group established to answer redirected calls. In addition to redirecting a call to a local answering position, Call Coverage can redirect calls based on time-of-day, redirect calls to a remote location, and allow users to change back and forth between two lead-coverage paths from either an on- or off-site location.

Call Detail Report (CDR)—A CDR records on disk or paper the details of incoming and outgoing telephone calls, including source and destination, time of day, and call length.

Call Management System (CMS)—CMS is an adjunct (basic software package or optional enhanced software package) that collects call data from a switch resident ACD. CMS provides call management performance recording and reporting. It can also be used to perform some ACD administration. CMS allows users to determine how well their customers are being served and how efficient their call management operation is.

Call Prompting—Call Prompting is a call management method that uses specialized call vector commands to provide flexible handling of incoming calls based on information collected from a caller. One example is when a caller receives an announcement and is then prompted to select (by means of dialed number selection) a department or option listed in the announcement.

Call Treatment—Call treatment refers to a call center's automatic call-handling procedure. Each call treatment is programmed into the ACD, using a simple scripting language. Most call treatments include elements such as waiting time, announcements, on-hold music, and telephone system instructions. Separate treatments are normally programmed for office hours, after-hours, or special events.

Call Usage Rate (CUR)—CUR is the time a PBX spends processing calls. See PUR.

Call Vectoring—Call Vectoring is an optional software package that allows processing of incoming calls according to a programmed set of commands. Call Vectoring provides a flexible service allowing direct calls to specific and/or unique call treatments.

Caller ID—Caller ID is a service that provides calling party information on a standard telephone line. This information is provided by means of a specified modem protocol between ringing signals.

Centi Call Seconds (CCS)—CCS also stands for hundred call seconds. One CCS is equivalent to 100 seconds holding time on any number of trunks. Thirty-six CCS is equivalent to one Erlang.

Center Stage Switch (CSS)—A CSS is the central interface between the PPN and EPNs. The CSS serves as a hub for WAN carriers that connect multiple locations to the same central switch.

Central Office (CO)—A CO is the telephone facility where telephone users' lines (local loops) are joined to switching equipment that connects telephone users to each other.

Centralized Attendant Service (CAS)—CAS is a system feature used when more than one switch is employed. CAS is an attendant or group of attendants that handles the calls for all switches in a particular network.

Cladding—Cladding is the clear plastic or glass layer that encloses the light-transmitting core of a fiber optic cable. The cladding has a lower refractive index than the core, and reflects the light signal back into the core as the light propagates down the fiber.

Class of Service (CoS)—CoS provides preferential treatment to certain data types on the network. IEEE 802.1p allows administrators to define up to eight separate traffic CoSs, each controlling the type of service each frame experiences on the network.

Cloud—Any switched network that provides service while hiding its functional details from its users is referred to as a cloud. A user simply connects to the edge of the cloud, and trusts the network to handle the details of moving a signal or data across to its destination. The public-switched telephone system and Internet are two well-known examples of cloud networks.

Co-location—A physical and business arrangement to connect the network of a CLEC to that of the ILEC is referred to as co-location. To do this, a CLEC usually installs interconnection equipment at the ILEC's central switching office.

Committed Information Rate (CIR)—CIR is the guaranteed average data rate for a frame relay service.

Common Carrier—A common carrier is a company that must offer its services to all customers at the prices and conditions outlined in a public tariff.

Common Channel Signaling (CCS)—CCS dedicates a separate communications channel to control signaling, which eliminates problems associated with control signaling within a voice channel. CCS has culminated with ITU-T SS7, also called SS #7, which will gradually be adopted by most networks.

Commutator—An electronic device that converts alternating current to direct current is referred to as a commutator.

Compact Modular Cabinet (CMC1) Media Gateway—A CMCI is an Avaya media gateway that consists of a floor or wall mount compact cabinet containing control and interface circuit packs. A CMC1 can support the DEFINITY server CSI and the S8100/8500/8700 media servers.

Competitive Access Provider (CAP)—A CAP is a company that provides fiber optic links to connect urban business customers to IXCs, bypassing the LEC. Once these fiber optic links are in place in major metropolitan areas, CAPs will begin to expand their service offerings.

Competitive Local Exchange Carrier (CLEC)—CLECs are telecommunications resellers, or brokers, who sell data services, Internet access, and local toll calling to businesses and residential customers. Some CLECs route calls over a mix of their own fiber optic, wireless, and copper lines, as well as over facilities they buy at a discount from LECs.

Computer Telephony Integration (CTI)—CTI represents a variety of services made possible by connecting a telephone switch, such as a PBX, to a computer system. For example, a computer can control telephone switching and call routing. Alternately, a telephone switch can pass the identity of incoming callers to the computer system.

Constant Bit Rate (CBR)—CBR information requires synchronization between sender and receiver and a specified bandwidth to make sure information is communicated accurately. CBR service is used by voice, video, and similar time-sensitive traffic. The term CBR is typically associated with protocols designed for handling multimedia traffic, such as ATM.

Converged Network—A converged network is a network that supports both data and voice communication, or data and multimedia.

Copper Pair—The term copper pair refers to two copper wires that carry voice or data signals to a customer. See local loop.

Core—The core is the innermost layer of a fiber optic cable, made of clear glass or plastic. The core carries light signals down the fiber.

Coverage Answer Group—A coverage answer group is a group of up to eight voice terminals that ring simultaneously when a call is redirected to it by Call Coverage. Any one of the group can answer the call.

Coverage Call—A coverage call is a call that is automatically redirected from the called party's extension to an alternate answering position when certain coverage criteria are met.

Coverage Path—The order in which calls are redirected to alternate answering positions is referred to as the coverage path.

Cyclic Redundancy Check (CRC)—CRC is the mathematical process used to check the accuracy of data being transmitted across a network. Before transmitting a block of data, the sending station performs a calculation on the data block and appends the resulting value to the end of the block. The receiving station takes the data and CRC value, and performs the same calculation to check the accuracy of the data.

Data Service Unit/Channel Service Unit (DSU/CSU)—A DSU/CSU is the hardware required to connect a common carrier connection (leased line) to a router. A DSU takes information from a LAN device and creates digital information suitable for public transmission facilities. A CSU is the device that actually generates the transmission signals on the local loop (telephone channel). DSUs are normally coupled with CSUs in one device called a DSU/CSU.

DCS+—DCS+ is an Avaya proprietary signaling protocol that enhances DCS by passing control information over public networks, such as ISDN-PRI. DCS+ signaling travels between DCS nodes over the ISDN-PRI D-channel.

DEFINITY Enterprise Communications Server (ECS)—DEFINITY ECS is a digital switch that processes both voice and data traffic.

Dialed Number Identification Service (DNIS)—DNIS is a feature of toll-free service (800/888/877) that sends the dialed digits to the called destination. This can be used with a display voice terminal to indicate the type of call to an agent. For example, the destination number can classify a call or caller, depending on the product or service the destination number is associated with.

Digital Access Cross-Connect Switch (DACS)—DACS is a connection system that establishes semipermanent (not switched) paths for voice or data signals. All physical wires are attached to a DACS once, then electronic connections between them are made by entering instructions.

Direct Inward Dialing (DID)—DID is a process by which a PBX routes calls directly to a particular extension (identified by the last four digits). Incoming trunks must be specifically configured to support DID.

Direct Inward System Access (DISA)—DISA is a PBX feature that allows an outside caller to dial directly into the PBX system, then access the system's features and facilities remotely. DISA is typically used to allow employees to make long-distance calls from home or any remote area, using the company's less expensive long-distance service. To use DISA, an employee calls a special access number (usually toll-free), then enters a short password code.

Divestiture—The breakup of AT&T and the Bell System by the U.S. Justice Department in 1984 is an example of a divestiture. To end an illegal monopoly, AT&T was ordered to separate itself from its 22 local Bell operating companies, which were reorganized into seven RBOCs. AT&T was then restricted to the long distance business, while the RBOCs were limited to local (intraLATA) service. See RBOC.

Dumb Terminal—A terminal that totally depends on a host computer for processing capabilities is referred to as a "dumb" terminal. Dumb terminals typically do not have a processor, hard drive, or floppy drives; only a keyboard, monitor, and method of communicating to a host (usually through some type of controller).

Dynamic Random Access Memory (DRAM)—DRAM is the memory of a computer that can be read from or written to by computer hardware components such as a CPU. DRAM is volatile memory; data is erased from DRAM when a computer loses power or shuts down.

E1—E standards are the European standards that are similar to the North American T-carrier standards. E1 is similar to T1.

E&M—The letters "E&M" are derived from the words ear and mouth; the "mouth" wire of a loop transmits supervisory signals, and the "ear" wire receives them. An E&M trunk uses two pairs of wires (one E pair and one M pair) instead of the single pair used on today's trunks. Some older PBX systems are configured to use two-pair E&M trunks as tie lines (sometimes called tie trunks).

Electronic Tandem Network—A private wide area telephone network that uses leased lines to link multiple PBX systems is referred to as an electronic tandem network. Each PBX can serve as a tandem switch, routing calls to any other PBX. The tandem network functions as one unified telephone system.

Erlang—One Erlang equals one hour of holding time on any number of trunks. One Erlang is equivalent to 36 CCS. This unit of measurement was named for A.K. Erlang, a Danish telephone engineer who developed a method of estimating the capacity of telephone switches and trunk groups.

Erlang B, C—Erlang B and C are traffic engineering formulas that determine the number of trunks required to serve a particular volume of call traffic at a given grade of service. Erlang B assumes all blocked calls disappear, to redial later or not at all. Erlang C assumes all blocked calls enter a queue, and wait indefinitely until served.

Expansion Port Network (EPN)—An EPN contains additional ports that increase the number of connections to trunks and lines. Unlike a PPN, an EPN does not contain an SPE.

Expert Agent Selection (EAS)—EAS is an optional feature available, with Generic 3 and Generic 2.2, that uses Call Vectoring and ACD in the switch to route incoming calls to the correct agent, on the first try, based on skills.

Extension—Voice terminals connected to a PBX/switch by means of telephone lines are referred to as extensions. The term also defines the three-, four-, or five-digit numbers used to identify the voice terminal to the PBX/switch software for call routing purposes.

Fiber Optic—Fiber optic is a thin strand of glass or plastic that transmits a light beam by bending it so that it remains contained within the strand. By using fiber optic transmission, digital signals can travel long distances with a high degree of accuracy.

Flash Read Only Memory (ROM)—Flash ROM is non-volatile computer memory that can be erased or reprogrammed; however, it retains its data when power is lost or a computer shuts down.

Foreign Exchange (FX)—An FX is a trunk service that lets businesses in one city operate in another city by allowing customers to call a local number. The number is connected, by means of a private line, to a telephone number in a distant city.

Fractional T1 (FT1)—FT1 is a leased-line service that provides data rates from 64 Kbps to 1.544 Mbps, by allowing a user to purchase one or more channels of a T1 link. If a customer needs less bandwidth than 1.544 Mbps, FT1 is a low-cost alternative to purchasing a full T1.

Frame Relay—Frame relay is a packet-forwarding WAN protocol that normally operates at speeds of 56 Kbps to 1.5 Mbps.

G.711—G.711 is one of a series of ITU-T voice digitizing algorithms. G.711 transfers digitized audio at 48, 56, and 64 Kbps.

G.723—G.723 is one of a series of ITU-T voice digitizing algorithms. G.723 transfers digitized audio at 5.3 or 6.3 Kbps.

G.729—G.729 is one of a series of ITU-T voice digitizing algorithms. G.729 transfers digitized audio at 8 Kbps.

Grade of Service—The probability that an incoming call will be blocked during the busy hour is referred to as the grade of service. For example, a grade of service of 0.001 means that, on average, one call in one thousand will be blocked (receive a busy signal).

H.323—H.323 is the ITU-T recommended set of standards and protocols used to support multimedia conferencing on packet-based, best effort networks, such as the Internet and IP LANs. Included in the recommendation are several protocols that control call setup and teardown and voice and video signal conversion and compression.

Hertz (Hz)—Radio signals are measured in cycles per second, or Hz. One Hz is 1 cycle per second; 1,000 cycles per second is 1 kHz; 1 million cycles per second is 1 MHz.

High-Level Data Link Control (HDLC)—The HDLC protocol suite represents a wide variety of link layer protocols such as SDLC, LAPB, and LAPD.

High Speed Serial Interface (HSSI)—HSSI is a serial data interface that can operate at speeds up to 52 Mbps.

Holding Time—The holding time is the total time a call uses, or "holds," a trunk.

Hunt Group—A group of trunks/agents selected to work together to provide specific routing of special purpose calls is referred to as a hunt group.

Hybrid Fiber-Coax (HFC)—HFC is a network design method, common in the cable television industry, that combines optical fiber and coaxial cable into a single network. Fiber optic cables run from a central site to neighborhood hubs. From the hubs, coaxial cable serves individual homes.

Incumbent Local Exchange Carrier (ILEC)—An ILEC is the same as a LEC or RBOC.

Induction Heater—An induction heater is a device that generates heat by placing a heat-conducting material in a fluctuating electromagnetic field. The changing field induces a circulating flow of electrical energy in the conductor, which produces heat.

Integrated Services Digital Network (ISDN)—ISDN is a digital multiplexing technology that can transmit voice, data, and other forms of communication simultaneously over a single local loop. ISDN-BRI provides two "bearer" channels (B channels) of 64 Kbps each, plus one control channel (D channel) of 16 Kbps. ISDN-PRI is also called T1 service. It offers 23 B channels of 64 Kbps each, plus 1 D channel of 64 Kbps.

Interactive Voice Response (IVR)—IVR is an interface technology that allows outside callers to control a computer application and input information using their telephone keypads. All IVRs can speak back the results of the computer application, and some can also be programmed to fax back the results.

Interexchange Carrier (IXC)—An IXC is a long distance company (such as AT&T or MCI) that provides telephone and data services between LATAs.

Interference—Any energy that interferes with the clear reception of a signal is referred to as interference. For example, if one person is speaking, the sound of a second person's voice interferes with the first. See noise.

Interflow—Interflow refers to the redirection of a call to a destination outside the local switch network (different switch system). Interflow is used when a split's/skill's queue is heavily loaded, or a call arrives after normal work hours.

International Telecommunication Union-Telecommunications Standardization Sector (ITU-T)—ITU-T is an intergovernmental organization that develops and adopts international telecommunications standards and treaties. ITU was founded in 1865 and became a United Nations agency in 1947.

Internet Protocol (IP) Telephony—IP Telephony refers to voice telephone service provided over an IP network instead of the public-switched telephone service.

Interoperable—Systems that can work together are considered interoperable. To ensure interoperability, hardware and software manufacturers develop common standards to define the way devices connect and programs exchange information.

Intraflow—Redirection of a call to a destination within the local switch network (same switch system) is referred to as intraflow. Intraflow is used when a split's/skill's queue is heavily loaded or a call arrives after normal work hours.

ITU-T V.92—ITU-T V.92 is dial-up modem specification that features quick connect, modem-on-hold and PCM Upstream. The quick connect feature shortens the modem handshaking portion of the call's setup procedures. Modem-on-hold allows a caller to put a modem call on-hold to pick up an inbound voice call (used in conjunction with call waiting on the modem telephone line). PCM upstream promises upstream speeds of up to 48Kbps.

Key System—A key system is a simple business telephone system that provides multiple inside extensions access to any of several incoming lines.

Leave Word Calling—Leave word calling is a system feature that allows messages to be stored for any ACD split/skill, and allows for retrieval by a covering user of that split/skill or a system-wide message retriever.

Line—A line is a transmission path carried over a physical local loop, but not the same thing as the loop itself. It is a path that connects a telephone switch to an individual user.

Link Access Procedure Balanced (LAPB)—LAPB (or LAP-B) is an HDLC protocol subset used primarily in X.25 communications.

Link Access Procedure for D Channel (LAPD)—LAPD (or LAP-D) is part of the ISDN layered protocol. It is very similar to LAPB. LAPD defines the protocol used on the D channel to interface with a telephone company's SS7 network for setting up calls and other signaling functions.

Linux—Linux is an open-source (freeware) version of UNIX created as a class project by a Finnish graduate student named Linux Torvalds.

Local Access and Transport Area (LATA)—LATAs are geographic calling areas within which an RBOC may provide local and long distance services. LATA boundaries, for the most part, fall within states and do not cross state lines. Each LATA is identified by a unique area code. Calls that begin and end within the same LATA (intraLATA) are generally the sole responsibility of the local telephone company (LEC), while calls that cross to another LATA (interLATA) are passed on to an IXC.

Local Exchange—A geographic region and group of subscribers served by a single CO is referred to as a local exchange.

Local Exchange Carrier (LEC)—A LEC is a company that makes telephone connections to subscribers' homes and businesses, provides telephone services, and collects fees for those services. The terms LEC, ILEC, and RBOC are equivalent.

Local Loop—The pair of copper wires that connects a customer's telephone to the LEC's CO switching system is referred to as the local loop.

Local Survivable Processor (LSP)—An LSP is an Avaya media server co-located with a remote G700 media gateway that operates as a hot spare in the remote location in the case that the media gateway loses communications with the primary call processor. In the instance of a communication failure between the media gateway and the primary call processor, the LSP takes over call processing for the local site. If the link between the primary and remote site caused the original failover to the LSP, the remote site will not be able to place calls to the primary site over the failed link until the link is restored to service.

Look Ahead Interflow (LAI)—LAI is an ACD feature that allows a busy contact center to move some or all calls to another ACD better able to service them. Vectors define call conditions that can trigger the ACD to send the call to another site, and the receiving site can choose to accept or reject the call.

The call remains in queue at the sending location while it determines whether to forward the call on or hold it for the next available agent.

Management Information Systems (MIS)—MIS is the traditional name of the department responsible for a company network or computing infrastructure.

Microwave—Microwaves are high-frequency radio waves, commonly used for wireless telephone transmission. While broadcast radio stations usually transmit between 535 and 1,605 kHz, cellular phone systems operate in bands of 824 to 849 MHz and 869 to 894 MHz. See hertz.

Modified Final Judgement (MFJ)—In the context of this course, MFJ refers to the court decision, effective January 1, 1984, that required AT&T to split apart and divest itself of its 22 local telephone companies.

Molina—The term Molina is a name sometimes used for the Poisson traffic engineering formula and tables. The traffic formula and associated tables were actually developed by E.C. Molina. However, he gave full credit to S.D. Poisson, who had discovered the mathematical principle in the 1820s.

Multicarrier Cabinet (MCC1) media gateway—An MCC1 is a 70 inch cabinet that houses up to five carriers, a fan unit, and a power distribution unit. Each carrier contains control, interface, port, and/or service circuit packs. An MCC1 can be a PN, EPN, or an auxiliary cabinet.

Multiplexer (MUX)—MUX refers to computer equipment that allows multiple signals to travel over a single channel. Multiple signals are fed into a MUX and combined to form one output stream.

National Electrical Code (NEC)—The NEC is a set of safety standards and rules for the design and installation of electrical circuits, including network and telephone cabling. The NEC was developed by a committee of ANSI, and has been adopted as law by many states and cities. Versions of the NEC are dated by year; each local area may require compliance with a different version of the NEC.

Network Control/Packet Interface (NetPkt)—The NetPkt board connects the process and port circuit packs to the TDM bus and provides an interface to the processor for D-channel signaling over the packet bus. The NetPkt board replaces both the NETCON and PACCON boards in newer systems.

Network Controller (NETCON)—The NETCON board is used in older systems; it connects the process and port circuit packs to the TDM bus.

Next Hop Resolution Protocol (NHRP)—NHRP is used by routers to dynamically discover the MAC address of other routers and hosts connected to an NBMA network. These systems then can communicate directly without requiring traffic to use an intermediate hop, increasing performance in ATM, frame relay, SMDS, and X.25 environments.

Night Service—Night service is used when a call arrives after normal work hours. The call can be redirected to another destination, such as another split/ skill, an extension, the attendant, an announcement with forced disconnect, or a message center. Night Service can take one of three forms:

- Hunt Group (Split/Skill) Night Service

- Trunk Group Night Service

- System Night Service

Node—A switching or control point for a network is referred to as a node. Nodes are either tandem (they receive signals and pass them on) or terminal (they originate or terminate a transmission path).

Noise—Any undesired signal or signal distortion is referred to as noise. Noise is often caused by some kind of interference. See interference.

Non-Broadcast Multi-Access (NBMA)—NBMA describes a multi-access network that either does not support broadcasting (X.25, for example) or in which broadcasting is not feasible, such as an SMDS broadcast group or an extended Ethernet that is too large.

Non-Facility Associated Signaling (NFAS)—NFAS is a form of ISDN-PRI out-of-band signaling that allows a single D-channel to control multiple PRIs. This allows all but one of the physical T1 carriers to support 24 ISDN B-channels, requiring that only one PRI dedicate a signaling channel.

Nonvolatile Memory—Nonvolatile memory is computer memory that retains its data when power is lost or a computer is shut down. See flash ROM.

Occupational Safety and Health Administration (OSHA)—OSHA is the federal agency that establishes and enforces measurable standards for workplace safety.

Optical Carrier (OC)—OC is one of the optical signal standards defined by the SONET digital signal hierarchy. The basic building block of SONET is the STS-1 51.84-Mbps signal, chosen to accommodate a DS3 signal. The hierarchy is defined up to STS-48 (48 STS-1 channels), for a total of 2,488.32 Mbps capable of carrying 32,256 voice circuits. The STS designation refers to the interface for electrical signals. The corresponding optical signal standards are designated OC-1, OC-2, etc.

Outcalling—Outcalling is a voice mail feature that dials an outside number (such as a pager), that alerts a traveling user to new voice mail.

Packet Bus—The packet bus runs internally throughout each PN, and terminates on each end. It is an 18-bit parallel bus that carries logical links and control messages from the SPE, through port circuits, to endpoints, such as terminals and adjuncts. The packet bus carries X.25 links, SNIs, and remote management terminal traffic.

Packet Controller (PACCON)—The PACCON board is used in older systems; it provides an interface to the processor for D-channel signaling over the packet bus.

Packet Gateway (PGATE)—The PGATE board is used in newer systems; it connects the processor to the packet bus and terminates X.25 signaling.

Path Replacement—Path replacement is the process of rerouting an established call over a newer, more efficient path. The old call is then torn down, freeing up the resource. Path replacement is used on Q.SIG and DCS+ calls in conjunction with Look-Ahead Interflow (LAI) and Best Service Routing (BSR) to attempt to find the optimum path for an inbound call.

Path Replacement with Path Retention—When a call is transferred within a private network, the call's connection between the switches can be replaced with new connections while the call is active.

Permanent Virtual Circuit (PVC)—A PVC is a connection across a frame relay network, or cell-switching network such as ATM. A PVC behaves like a dedicated line between source and destination endpoints. When activated, a PVC will always establish a path between these two end points.

Personal Computer Memory Card International Association (PCMCIA)—The PCMCIA slot in a laptop was designed for PC memory expansion. NICs and modems can attach to a laptop through the PCMCIA slot.

Phase—A phase is a description of one wave's position relative to another at a particular point in time. Phase differences are measured in degrees from 0 to 180.

Point-to-Point Protocol (PPP)—PPP is a protocol that allows a computer to use TCP/IP by the means of a point-to-point link. PPP is based on the HDLC standard that deals with LAN and WAN links, and operates at the Data Link Layer of the OSI model.

Poisson—Poisson is the traffic engineering formula that determines the number of trunks required to serve a particular volume of call traffic at a given grade of service. Poisson assumes all blocked calls immediately redial, and keep redialing until they get through. The formula is named for S.D. Poisson, who first developed the mathematical approach in the 1820s; it is also called the Molina formula.

Port Network (PN)—PNs include both EPNs and PPNs. These two PNs differ in that the PPN contains an SPE. Both types of PNs consist of the following components: TDM bus, packet bus, port circuits, and interface circuits.

Port Usage Rate (PUR)—PUR is the number of PBX ports required to process a call. PUR is typically twice the CUR, because a simple call requires a minimum of two ports (input and output). See CUR.

Priority Queue—The priority queue is a segment of a split's/skill's queue from which calls are taken first.

Private Branch Exchange (PBX)—A PBX is a sophisticated business telephone system that provides all the switching features of a telephone company's CO switch. Today's PBXs are fully digital, not only offering very sophisticated voice services, such as voice messaging, but also integrating voice and data.

Private Signaling System Number 1 (PSS1 or Q.SIG)—PSS1 is an ISO standard that defines the ISDN signaling and control methods used to link PBXs in private ISDN networks. The standard extends the "Q" point in the ISDN logical reference model, which was established by ITU-T in its Q.93x series of recommendations that defined the basic functions of ISDN switching systems. Q.SIG signaling allows certain ISDN features to work in a single- or multivendor network.

Processor Interface (PI)—A PI packet gateway (PGATE) circuit pack supports X.25 signaling between DEFINITY systems and adjuncts. The PI circuit pack is not included in new DEFINITY systems; however, PI channels are common in existing installations that still use X.25 connections.

Processor Port Network (PPN)—A PPN is required by each DEFINITY system; it contains the switch processing element (SPE), the system memory, the packet controller, and the network controller circuit packs.

Provisioning—The process of allocating transmission lines, switching capacity, and central programming to provide telecommunications service to a customer is referred to as provisioning.

Pulse Code Modulation (PCM)—PCM is a method of converting an analog voice signal to a digital signal that can be translated accurately back into a voice

signal after transmission. A codec samples the voice signal 8,000 times per second, then converts each sample to a binary number that expresses the amplitude and frequency of the sample in a very compact form. These binary numbers are then transmitted to the destination. The receiving codec reverses the process, using the stream of binary numbers to recreate the original analog wave form of the voice.

Q.SIG—Also known as Private Signaling System Number 1 (PSS1), Q.SIG is an ISO standard that defines the ISDN signaling and control methods used to link PBXs in private ISDN networks. The standard extends the "Q" point in the ISDN logical reference model, which was established by the ITU-T in its Q.93x series of recommendations that defined the basic functions of ISDN switching systems. Q.SIG signaling allows certain ISDN features to work in a single- or multi-vendor network.

Queue—A queue is a collection point where calls are held until an agent or attendant can answer them. Calls are ordered as they arrive and are served in that order. Depending on the time delay in answering the call, announcements, music, or prepared messages may be employed until the call is answered.

Queue Directory Number (QDN)—QDN is an associated extension number of a split. It is not normally dialed to reach a split. The split can be accessed by dialing the QDN. The QDN is also referred to as a split group extension.

Reduced Instruction Set Computer (RISC)—The term RISC refers to microprocessors that contain fewer instructions than traditional CISC processors, such as Intel or Motorola. As a result, they are significantly faster. RISC processors have been used in most technical workstations for some time, and a growing number of PC-class products are based on RISC processors.

Regional Bell Operating Company (RBOC)—An RBOC is one of seven companies formed from AT&T's 22 local telephone companies during the breakup of the Bell System. The terms RBOC, LEC, and ILEC are equivalent. The original seven RBOCs were:

* Ameritech
* Bell Atlantic
* Bellsouth
* New York New England Telephone Company (NYNEX)
* Pacific Telesis
* Southwestern Bell Communications

- U S WEST

Request for Comment (RFC)—An RFC is one of the working documents of the Internet research and development community. A document in this series may be on essentially any topic related to computer communication, from a meeting report to the specification of a standard.

Request for Proposal (RFP)—An RFP is a formal document that specifies the equipment, software, and services an organization wants to buy, and asks vendors to submit written bids offering their best prices.

Router—A router is a Layer 3 device, with several ports that can each connect to a network or another router. A router examines the logical network address of each packet, then uses its internal routing table to forward the packet to the routing port associated with the best path to the packet's destination. If the packet is addressed to a network that is not connected to the router, the router will forward the packet to another router that is closer to the final destination. Each router, in turn, evaluates each packet, then either delivers the packet or forwards it to another router.

Sampling—Sampling is the scientific technique of describing the characteristics of a population by examining only a portion of the population.

Service Observing—Service observing is a feature used to train new agents and observe in-progress calls. The observer (split/skill supervisor) can toggle between a listen-only mode or listen/talk mode during calls in progress.

Single Carrier Cabinet (SCC1) media gateway—An SCC1 consists of a single carrier containing the processor or IPSI circuit packs, tone-clock, port and service circuit packs, and a power supply. Up to four SCC1s may be stacked to form a PN.

Signaling System 7 (SS7)—SS7 (also called SS #7) is an out-of-band system that exchanges control signals and call routing information between CO switches. It is a separate network that connects all COs, regardless of where they are or to whom they belong.

Skill—Skill refers to the ability assigned to an agent to meet a specific customer requirement or call center business requirement.

Span—A digital connection between a CO and terminal switch, such as a PBX, is referred to as a span.

Split—A split is a group of extensions/agents that can receive standard and/or special purpose calls from one or more trunk groups. Each split can be served by its own queue. If splits are defined according to

agents' knowledge, they are called "skills." Depending on the ACD software, an agent can be a member of multiple splits/skills.

Split/Skill Administration—The split/skill administration refers to the ability to assign, monitor, or move agents to specific splits/skills. It involves changing reporting parameters within the system.

Split/Skill Supervisor—A split/skill supervisor is a person assigned to monitor/manage each split/skill and queue to accomplish specific objectives. A supervisor can assist agents on ACD calls, be involved in agent training, and control call intra-/interflow.

Spread—The range between the largest and smallest values in a scientific sample is referred to as the spread. When the range of a sample is close to its average, the sample is more likely to accurately represent the total population being studied.

Statistical Multiplexing—Statistical multiplexing, or STDM, is a more flexible method of TDM. TDM allocates a fixed number of time slots to each channel, regardless of whether the channel has data to send. In contrast, a statistical multiplexer analyzes transmission patterns to predict "gaps" in a channel's traffic that can be temporarily filled with part of the traffic from another channel.

Subscriber Line Carrier (SLC)—SLC is a method of using T1 multiplexing technology to carry more lines over existing wires.

Subscriber Network Interface (SNI)—The SMDS protocol specifies how to connect CPE with an SMDS network. The point at which the CPE interfaces with the SMDS network is the Subscriber Network Interface (SNI). This interface, connecting a customer to the SMDS cloud, is usually implemented by means of T1 or T3 lines.

Switch Node (SN)—An SN reduces the amount of interconnect cabling between the PPN and EPNs by acting as a hub to distribute cabling. A system using a CSS can connect 3 to 43 PNs. A CSS can consist of up to three SN carriers.

Switch Processing Element (SPE)—When a telephone goes off-hook, or another type of device signals call initiation, the SPE receives a signal from the port circuit connected to the device. The SPE collects the digits of the called number, and the switch is set up to make a connection between the calling and called devices.

Switched Multimegabit Data Service (SMDS)—SMDS is a connectionless service used to connect LANs, MANs, and WANs at rates up to 45 Mbps. SMDS is cell-oriented and uses the same format as the ITU-T B-ISDN standards. The internal SMDS protocols are called SIP-1, SIP-2, and SIP-3. They are a subset of the IEEE 802.6 standard for MANs, also known as DQDB.

Switched Virtual Circuit (SVC)—An SVC is a temporary connection established through a switched network. During data transmission, an SVC behaves like a wire between the sender and receiver. ATM VC and telephone connections are both examples of SVCs.

System Management Terminal—A system management terminal is the terminal from which system administration and maintenance is performed.

T1, T3—T1 and T3 are two services of a hierarchical system for multiplexing digitized voice signals. The first T-carrier was installed in 1962 by the Bell System. The T-carrier family of systems now includes T1, T1C, T1D, T2, T3, and T4 (and their European counterparts E1, E2, etc.). T1 and its successors were designed to multiplex voice communications. Therefore, T1 was designed such that each channel carries a digitized representation of an analog signal that has a bandwidth of 4,000 Hz. It turns out that 64 Kbps is required to digitize a 4,000-Hz voice signal. Current digitization technology has reduced that requirement to 32 Kbps or less; however, a T-carrier channel is still 64 Kbps. A T1 line offers bandwidth of 1.544 Mbps; a T3 offers 44.736 Mbps.

T-span—A T-carrier that connects a PBX to a CO is referred to as a T-span.

Tandem Switch—A telephone switch that forwards traffic between other switches is referred to as a tandem switch.

Threshold—A threshold is a point in time, or a criterion, that determines a certain action by a system. For example, the number of calls in a queue, or the time calls spend in a queue, can determine specific call treatments.

Tie Line, Tie Trunk—A tie line (tie trunk) is a dedicated circuit that links two points without having to dial a telephone number. Many tie lines provide seamless background connections between business telephone systems.

Time-Division Multiplexing (TDM)—TDM is a multiplexing technology that transmits multiple signals over the same transmission link, by guaranteeing each signal a fixed time slot to use the transmission medium.

TN2302AP IP Media Processor (Medpro)—The TN2302AP Medpro circuit pack provides packetized audio processing for an IP-enabled DEFINITY server or media gateway. The Medpro circuit pack supports the ITU-T H.323v2 recommendation and performs jitter buffering, echo cancellation, silence suppression, and DTMF tone detection. It supports the ITU-T G.711, G.723, and G.719 audio codecs.

TN2312AP IP Server Interface (IPSI)—The TN2312AP IPSI circuit pack provides call control messaging between Avaya media servers and IP-enabled port networks (PNs).

Transcoding—Transcoding is the process where a device converts a signal from one format to another. For example, an H.323 gateway transcodes packetized voice to analog for transport across the PSTN.

Translations—Translations are a set of programming instructions, stored within a CO switch, that defines the functions and services available on a line or trunk. The programming process is also called "translations," because it effectively tells a switch how to interpret, or convert, signals from other switches.

Transport Carrier—A telecommunications provider or telephone company is referred to as a transport carrier.

Trunk—A trunk is a transmission path that connects two telephone switches: two COs, a CO and a business PBX, or two PBXs.

Trunk Group—A trunk group is a group of trunks that provide identical communications characteristics. Trunks within trunk groups can be used interchangeably between two communications systems or COs to provide multiaccess capability. An ACD has its own preassigned trunk groups.

Trunk State—The current status of a trunk is referred to as the trunk state.

Type of Service (ToS), Quality of Service (QoS)— Users of the Transport Layer specify QoS or ToS parameters as part of a request for a communications channel. QoS parameters define different levels of service based on the requirements of an application. For example, an interactive application that needs good response time would specify high QoS values for connection establishment delay, throughput, transit delay, and connection priority. However, a file transfer application needs reliable, error-free data transfer more than it needs a prompt connection, thus it would request high QoS parameters for residual error rate/probability.

Unbundled Service—Unbundled service refers to a communications channel leased to a CLEC by the ILEC. "Unbundled" means that the ILEC provides only the transmission service, while the CLEC provides management, provisioning, repair, and billing.

Uniform (Universal) Dial Plan—A uniform dial plan is a private network numbering system that assigns unique extension numbers to each user, regardless of location. If locations in different cities are linked in a private telephone network, a uniform dial plan allows employees in different cities to dial each other using only extension numbers.

Usage Sensitive Pricing (USP)—Usage sensitive pricing is a telecommunications service pricing practice where the carrier bills the customer by line usage rather than by a flat fee. Carriers charge by the call, based on call duration, time of day, and call distance.

Virtual Circuit (VC)—A VC is a communication path that appears to be a single circuit to the sending and receiving devices, even though the data may take varying routes between the source and destination nodes.

Virtual Private Network (VPN)—A connection over a shared network that behaves like a dedicated link is referred to as a VPN. VPNs are created using a technique called "tunneling," which transmits data packets across a public network, such as the Internet or other commercially available network, in a private "tunnel" that simulates a point-to-point connection. The tunnels of a VPN can be encrypted for additional security.

Wide Area Telecommunications Service (WATS)—WATS is a discounted long distance service. Customers can purchase incoming and outgoing WATS separately.

X.25—X.25 is a connectionless packet-switching network, public or private, typically built upon leased lines from public telephone networks. In the United States, X.25 is offered by most carriers. The X.25 interface lies at OSI Layer 3, rather than Layer 1. X.25 defines its own three-layer protocol stack, and provides data rates only up to 56 Kbps.

INDEX

Automatic
    Call Distributor (See ACD)
    Number Identification (ANI) 111
    Route Selection (ARS) 211
Automatic-in processing 224
Avaya
    Access Security Gateway (See ASG)
    ASG 218
    Communications Manager (See
      ACM)

B

Bandwidth 10, 61, 66, 261, 383
Best Service Routing (See BSR)
Bias 420, 445
Billing 27
B-ISDN 355
Blocked call 420
Bouncing busy hour 420
Bridge 244
British Thermal Unit (BTU) 112
BSR 228, 472
Bursting 384
Burst range 318, 384, 472
Bursty 318
Bus 112, 160
Busy hour
    bouncing 420, 451
    constant 451
    engineering 450
    studies 135
BX.25 protocol 112, 167, 472

C

Cable
    fiber optic 12
    modems 292
    modem vs. ASDL 294
    vault 244

Call
    accounting 142
    blocked 420
    center 112
    coverage 112
    Detail Reports (See CDR)
    distribution 226
    following 137
    handling 149
    Management System (CMS) 112
    prompting 113
    ringing 46
    routing 43, 130
    treatment 113
    Usage Rate (CUR) 436
    vectoring 113
Call-by-call service 339
Call Detail Report (See CDR)
Caller ID 10
Calling 43
CAP 11
Capacity 437
Carried load 423
Carrier 177
CAS 113
CBR 318, 356, 474
CCS 420, 428
CDR 10, 27, 135, 472
Center Stage Switch (See CSS)
Centi Call Seconds (See CCS)
Centralized Attendant Service
  (CAS) 113
Central Office (See CO)
Centrex 71 to 75
Channel
    bonding 340
    overview 61
CIR 318, 366, 367, 384, 385, 473
Circuit
    connections 25
    costs 279